The
Industrial Wood Energy Handbook

The Industrial Wood Energy Handbook

Prepared by

The Technology Applications Laboratory

of the
Georgia Institute of Technology
Engineering Experiment Station
Atlanta, Georgia

VAN NOSTRAND REINHOLD COMPANY
NEW YORK   CINCINNATI   TORONTO   LONDON   MELBOURNE

Library of Congress Catalog Card Number: 83-1218
ISBN: 0-442-22085-5

Manufactured in the United States of America

Published by Van Nostrand Reinhold Company Inc.
135 West 50th Street
New York, New York 10020

Van Nostrand Reinhold Company Limited
Molly Millars Lane
Wokingham, Berkshire RG11 2PY, England

Van Nostrand Reinhold
480 Latrobe Street
Melbourne, Victoria 3000, Australia

Macmillan of Canada
Division of Gage Publishing Limited
164 Commander Boulevard
Agincourt, Ontario M1S 3C7, Canada

15 14 13 12 11 10 9 8 7 6 5 4 3 2 1

Library of Congress Cataloging in Publication Data

Main entry under title:
The Industrial wood energy handbook.
   Includes index.
   1. Wood as fuel—Handbooks, manuals, etc.
I. Georgia Institute of Technology. Technology Applica-
tions Laboratory.
TP324.I5 1983   662'.8   83-1218
ISBN 0-442-22085-5

Disclaimer

The contents of this report are offered as guidance. The Georgia Institute of Technology and the Engineering Experiment Station, and all technical sources referenced herein do not (a) make any warranty or representation, expressed or implied, with respect to the accuracy, completeness, or usefulness of the information contained in this report, or that the use of any information, apparatus, method, or process disclosed in this report may not infringe on privately owned rights; (b) assume any liabilities with respect to the use of, or for damages resulting from the use of, any information, apparatus, method, or process disclosed in this report. This report does not reflect official views or policy of the above mentioned institutions. Mention of trade names or commerical products does not constitute endorsement or recommendation for use.

Preface

In June 1977, over 250 people attended a Georgia Tech/Georgia Forestry Commission co-sponsored wood energy conference in Atlanta. A majority of the attendees represented non-forest products industries from the State of Georgia. It became immediately obvious that industrial wood energy was a topic of interest to a wide cross-section of industry, architect-engineer consultants, equipment vendors, and wood-lot owners.

Over the succeeding four years (under the sponsorship of the United States Department of Energy, the State of Georgia, the Tennessee Valley Authority, the Coastal Plains Regional Commission, and non-forest products industries), the Wood Energy Systems Division at Georgia Tech has grown from two part-time engineers to twelve full-time research engineers. Beginning with broad, general resource studies, projects were completed involving research and development, individual plant wood feasibility studies, full-scale industrial demonstration projects, and numerous wood technology seminars.

This handbook was prepared during the second year of the United States Department of Energy grant, ''A State Demonstration Program in Wood Energy.'' Also accomplished under the demonstration phase of this project were the installation of a 13,500 lb steam/hr wood boiler at a North Georgia textile mill (supplanting natural gas), and a 60,000 lb steam/hr peanut hull and wood-fired boiler (with cogeneration capability) at an agricultural processing facility.

As the commercial feasibility of industrial wood energy has been clearly demonstrated, the focus of Georgia Tech's wood energy program is shifting towards research and development—in the areas of gasification, wood fuel standards, behavior of stored wood fuels, low Btu gas fired turbines, and solid fuels technology. The publication of this handbook is the culmination of the early stages of this effort. The equipment, technology, engineering expertise, and fuels are in place. Based on sound economics, it remains for industry to implement alternatives to traditional petroleum fuel sources.

Acknowledgments

The Industrial Wood Energy Handbook was written under Contract Number DE-FG05-79-ET 23076 which was funded by the U.S. Department of Energy through the State of Georgia Office of Energy Resources. The authors would like to acknowledge Dr. Beverly Berger of the U.S. Department of Energy and Mr. Mark Zwecker of the Georgia Office of Energy Resources for their guidance and support during the writing of this handbook and the performance of the contract.

Georgia Institute of Technology
Engineering Experiment Station
Technology Applications Laboratory
Wood Energy Systems Division

Credits

MAJOR CONTRIBUTORS

William S. Bulpitt, Chief
Wood Energy Systems Division — PROJECT DIRECTOR

Grant B. Curtis, Jr., P.E.
Senior Research Engineer — TASK DIRECTOR

Steven J. Drucker
Research Engineer — PROJECT ENGINEER/EDITOR

Michael L. Brown, P.E.
Research Engineer

Robert J. Didocha
Research Engineer

Thomas F. McGowan, P.E.
Senior Research Engineer

James L. Walsh, P.E.
Senior Research Engineer

Robert D. Atkins, P.E.
Research Engineer

Dr. Badarinath S. Dixit, P.E.
Senior Research Engineer

ADDITIONAL ASSISTANCE

John C. Adams, Jr.
Research Engineer

David E. Harris
Research Engineer

Anthony D. Jape
Research Engineer

Bryan Miller
Champion Paper
Courtland, AL

Joseph Saucier
U.S. Forest Service
Athens, GA

Dr. Arthur Shavit
Research Scientist

Michael S. Smith
Research Engineer

F. Dee Bryson
Research Technician

Douglas Davis
Co-Operative Student

William Hartrampf III
Co-Operative Student

David Lapin
Co-Operative Student

David Pugmire
Co-Operative Student

Joanne Bocek
Senior Secretary

Introduction

With increasing frequency, industrial wood energy systems demonstrate prompt paybacks and operating cost decreases for the industrial and commerical sectors. Thus, the decision to implement an industrial wood energy system is neither philosophic, political, or moral—it is economic.

This volume addresses the elements necessary for the planning, economic analysis, environmental compliance, design, and equipment selection of an industrial/commercial wood-fired steam plant—focusing on the size range of 10,000 to 50,000 lb steam/hr. The required technology is well known. The multitude of vendors with track records of successful equipment installations evidences the operational reliability of wood systems. Actual sizing of system elements (fans, pumps, augers, conveyors, fly ash collectors, piping, ductwork, etc.) differs little or not at all from traditional fossil fuel fired steam plants. Specific design data for these components is widely available from the literature, consulting engineers, fabricators, and vendors.

HANDBOOK ORGANIZATION

The handbook is organized into three sections. Section I describes theory, hardware, applicable rules and regulations, and good operating practice for the wood-fired steam plant. The reader, familiarized with the nuts and bolts of generating thermal energy from wood, is left with the question: ''Can this technology be of benefit to my plant?''

Section II addresses this question by conducting the reader through the logistic maze of options, alternatives, methods of payback analysis

"

and determination of fuel supplies. Go/no-go indicators are presented at each stage of the decision-making process.

The final section presents the case history of a textile products plant which converted its steam plant from natural gas and oil to wood residues and whole green tree chips. The reader can note that in the economic analysis, some government funding was received by the plant, and the economic analysis benefitted in kind. At the time of the feasibility study, moreover, oil was priced at $26.00/barrel and natural gas prices are close to decontrol. Thus, the continued price spiral of petroleum fuels should serve as red flags to those involved in management of thermal generating plants.

It is hoped that the information presented in the following pages will enable the plant engineer, engineering consultant, financial planner, and owner/operator to make informed decisions concerning industrial-scale wood energy installations. Fabricators and vendors who market wood equipment (and turnkey systems) to the industrial market should find in these pages much useful information. For the reader being exposed to this field on a first-time basis, a comprehensive glossary of technical terms is appended.

Steven J. Drucker
Handbook Editor

CONTENTS

Preface   vii
Credits   xi
Introduction   xiii

SECTION I

1. Wood Fuel Combustion Theory   3
2. Wood Combustion Equipment for Steam and Process Heat   13
3. Wood Fuel Storage and Handling   60
4. Cogeneration   88
5. Wood-Fired Boiler Emissions and Control   100
6. Environment and Safety (Rules, Regulations and Safe   119
     Practice)

SECTION II

7. Feasibility Study Methodology   133
8. Wood Fuel Supply and Purchasing   141
9. Economic Analysis for Wood Combustion Systems   145
10. Wood Fuel Processing Routes and Economics   152

SECTION III

11. Wood Fuel Processing Network   169
12. Case History–Integrated Products, Inc.   172

APPENDICES

1. Equipment Manufacturers/Vendors	187
2. State Forestry Commissions and USDA Forestry Service Offices (by State)	209
3. 1981 Wood Fuel Supply Survey—Georgia	213
4. Bibliogaphy and References	228
5. Glossary	232
Index	237

Section I

1

Wood Fuel Combustion Theory

INTRODUCTION

Energy recovery from wood is achieved by direct combustion or indirectly by thermochemical conversion. Direct combustion entails burning the solid wood. Indirect methods convert the wood to a liquid or gas. The wood-derived liquid or gaseous fuel is then burned to yield heat and combustion by-products. A discussion of types of equipment and combustion processes is presented in Chapter 2.

WOOD FUEL PROPERTIES

Wood fuel properties of concern during combustion include moisture content, particle size, proximate analysis, ultimate analysis, and heating value.

Moisture Content

Moisture content is described in one of two ways: (1) wet basis; and (2) dry basis. Those concerned with power generation most often consider moisture content on a wet basis. The wet basis moisture content directly reflects the fuel value of wood. Knowledge of both methods of calculating moisture content will be important when arranging wood fuel purchases—especially mill residues.

The moisture content (M.C.) of wood on the wet basis is the weight of water in a wood sample divided by the total weight of the sample:

$$\text{M.C. (wet basis) }\% = \frac{100 \times \text{weight of water}}{\text{weight of dry wood} + \text{weight of water}}$$

EXAMPLE: A one pound sample of wood is found to have 50% M.C. (wet basis). The results of laboratory analysis showed:

water	½ pound H_2O
wood	½ pound bone dry wood

$$M.C. = ½ \text{ lb}/(½ + ½) \text{ lb} \times 100 = 50\% \text{ wet basis}$$

The dry basis moisture content is favored by foresters and producer/manufacturers of wood products (a prime source of mill residues). The M.C. of wood on the dry basis is the fractional water content or the weight of water divided by the sample weight when dried:

$$\text{Moisture Content (dry basis) \%} = \frac{\text{weight of water}}{\text{weight of dry wood}} \times 100$$

Using the same example, a one pound sample which is half water and half bone dry wood by weight would have a wet weight of one pound, a bone dry weight of one half pound, and a M.C., dry basis, of 100%.

$$M.C. \text{ (dry basis) \%} = \frac{½ \text{ lb water}}{½ \text{ lb bone dry wood}} \times 100$$

To find wet basis from dry basis:

$$M.C. \text{ (wet basis) \%} = \frac{M.C. \text{ Dry}}{M.C. \text{ Dry} + 100} \times 100$$

To find dry basis from wet basis:

$$M.C. \text{ (dry basis) \%} = \frac{M.C. \text{ Wet}}{100 - M.C. \text{ Wet}} \times 100$$

Figure 1-1[40] illustrates graphically the relationship between wet basis and dry basis moisture contents. Table 1.1 covers the normally encountered range:

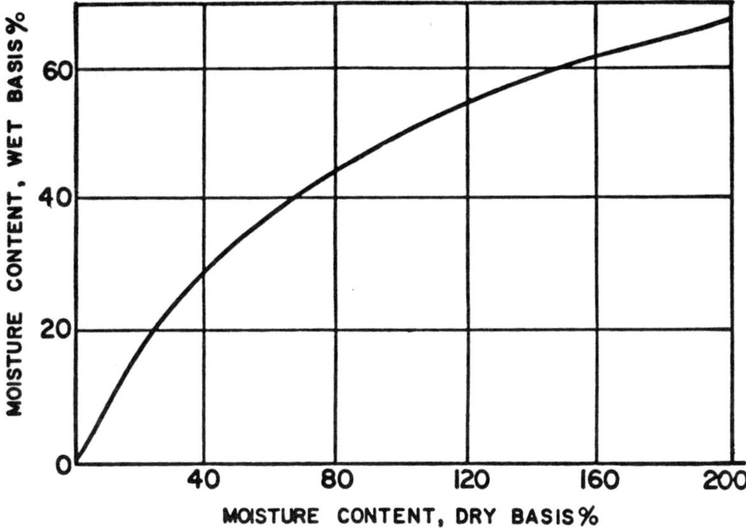

Figure 1-1. Moisture content comparison.

**TABLE 1-1. Moisture
Content: Comparison
Wet/Dry Bases.[40]**

WET BASIS (%)	DRY BASIS (%)
0	0
5	5
10	11
15	18
20	25
30	43
40	67
50	100
60	150

Unless noted, all moisture contents in this book are cited on a wet basis.

High moisture content fuelstock reduces combustion efficiency. The vaporization of water to steam requires a heat input of 1,000 Btu/lb of water. Energy which could otherwise be useful in steam production is thus diverted to drying the wood fuel in the combustion chamber prior

TABLE 1-2. Higher Heat Values of Some Wood Fuels.[40]

WOOD FUEL	MOISTURE CONTENT, WET BASIS (%)	HIGHER HEATING VALUE (BTU PER LB)
Whole tree chips	50	4,000
Dry planer shavings	13	6,960
Green sawdust	50	4,000
Dry sawdust	13	6,960
Wood pellets	10	7,200

to actual burning of the wood. Overall boiler capacity is also affected as illustrated in Figure 1-2.[66] Because the cost of equipment for pre-drying the fuel is high, the use of green wood fuels (M.C. between 50 to 65%) is often justifiable.

Particle Size

The particle size of wood fuel entering the combustion chamber, more specifically the surface area/mass ratio of the discrete pieces, will have great effect on combustion efficiency, boiler design, and quantity and method of introduction of combustion air. Improperly sized fuel may not burn completely, and valuable heat energy can be lost in the form of carbon-rich bottom ash and fly ash. Combustion units are designed specifically for certain size ranges of fuels, and close attention should be paid to maintaining equipment size requirements. Table 1-3 shows typical sizes of wood fuel (as received) for combustion.

TABLE 1-3. Typical Sizes of Wood Fuel— As Received.

TYPE	DIMENSION
Whole green tree chips	2 in. x 2 in. x ¼ in.
Sawdust	⅛ in. to ¼ in.
Mill residue	⅛ in. to several feet long
Pellets	¼ in. diameter x ½ in. long

Proximate Analysis

Proximate analysis describes the volatiles, fixed carbon, and ash present in a fuel as a percentage of dry fuel weight. The amounts of volatiles

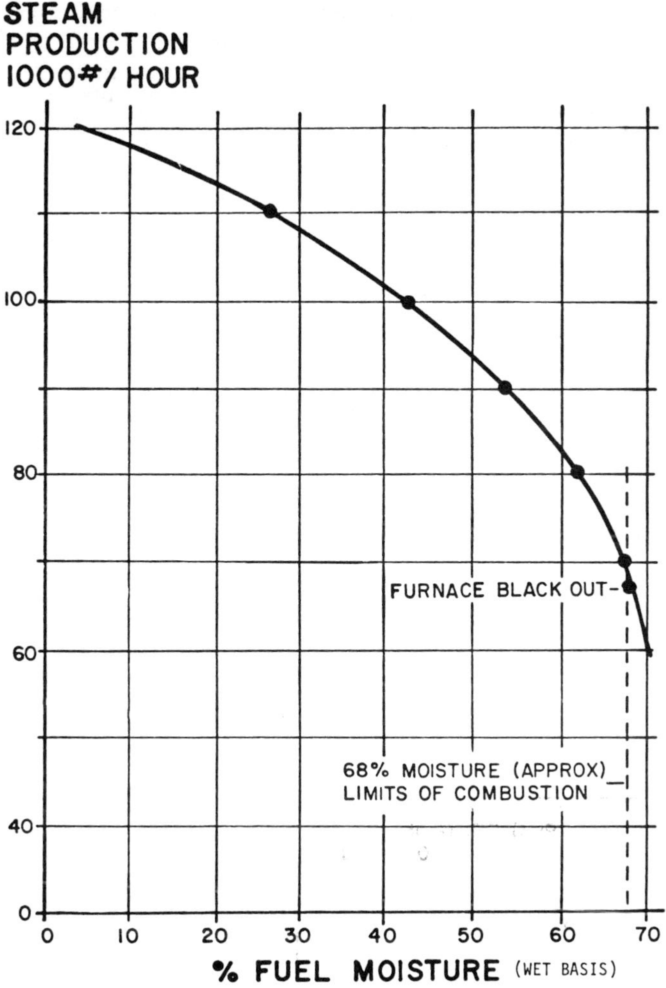

Figure 1-2. Boiler steam production vs. fuel moisture content.

and fixed carbon directly affect the heating value of the fuel, the flame temperature, and the process by which combustion is achieved. The ash content is important in the design of air pollution control equipment, boiler grates, and bottom ash handling equipment. From Table 1-4, it can be seen that wood has far lower ash content than coal, a similar amount of volatiles, and less fixed carbon. Ash contents in the range of 0.5 to 3% for wood fuels have also been noted in the literature.

TABLE 1-4. Proximate Analysis of Wood and Coal.[38]

	MOISTURE CONTENT (%)	VOLATILE MATTER (%)	FIXED CARBON (%)	ASH (%)
Bituminous coal	2.5	37.60	52.90	7.00
Hardwood (wet)	45.6	48.58	5.52	0.30
Hardwood (dry)	0	89.31	10.14	0.56
Southern Pine (wet)	52.3	31.50	15.90	0.29
Southern Pine (dry)	0	66.00	33.40	0.60

Ultimate Analysis

The ultimate analysis of a fuel describes its elemental composition as a percentage of the sample's dry weight.

Note that wood fuels are almost devoid of sulfur. Coupled with low ash content, these two properties make wood a highly desirable fuel from the standpoint of pollution control costs.

TABLE 1-5. Ultimate Analysis of Wood Species.[16]

WOOD SPECIES	HYDROGEN	CARBON	NITROGEN	OXYGEN	SULFUR	ASH
			(wood)			
California Redwood	5.9	53.5	0.1	40.3	trace	0.2
Western Hemlock	5.8	50.4	0.1	41.4	0.1	2.2
Douglas Fir	6.3	52.3	0.1	40.5	trace	0.8
Pine (sawdust)	6.3	51.8	0.1	41.3	trace	0.5
			(bark)			
Western Hemlock	6.2	53.0	0.0	39.3	trace	1.5
Douglas Fir	5.8	51.2	0.1	39.2	trace	3.7
Loblolly Pine	5.6	56.3	—	37.7	trace	0.4
Longleaf Pine	5.5	56.4	—	37.4	trace	0.7
Shortleaf Pine	5.6	57.2	0.4	36.1	trace	0.7
Slash Pine	5.4	56.2	0.4	37.3	trace	0.7

(PERCENT OF DRY FUEL WEIGHT)

Heating Value

The heating value of fuel is the measure of heat released during the complete combustion of the fuel (with oxygen): from reactants at a given reference temperature and pressure to products at the same reference temperature and pressure.

Analytically derived formulas have been developed for prediction of the Higher Heating Value (HHV) of coals. When compared with the wealth of knowledge generated concerning fossil fuels, interest in wood as a fuel has long been dormant. Exact calculations are available for all components of wood fuel which will oxidize; however, it is difficult to quantify the contribution of volatiles to heating value. Energy recovery from wood has stirred interest in this area, and it is expected that standards for analytical predictors of wood's higher heating value will appear shortly in the literature.

For purposes of thermal energy generation, the higher (or gross) heating value is defined in ASTM Standard D2015-77: "Standard Test Method for Gross Calorific Value of Solid Fuel by the Adiabatic Bomb Calorimeter."

Quoting from "Steam-Its Generation and Use" Babcock and Wilcox 37th Edition:

> Most commercial fuels contain hydrogen as one of the constituents, and water, H_2O, is formed as a product of combustion when the hydrogen is burned in air. This water may remain in the vapor state, or it may be condensed to the liquid state, giving a substantial difference in the heat value. In determining the heat given up by a unit of such fuel, two values may be reported—the high, or gross heat value and the low, or net, heat value. For the high, or gross, heat value, it is assumed that any water vapor formed by the burning of the hydrogen constituent is all condensed and cooled to the initial temperature in the calorimeter at the start of the test. The heat of vaporization is therefore present in the reported value. For the low, or net, heat value, is is assumed that none of the water vapor condenses and that all of the products of combustion remain in a gaseous state.[64]

By definition then:

- *Higher Heating Value (HHV) = Gross Heating Value (GHV)* is determined in the laboratory using an oxygen bomb calorimeter. HHV, expressed in units of heat released/unit weight (Btu/lb), is the heat content of a fuel (wood or other). It is always necessary, when speaking of HHV (or GHV), *to note the Moisture Content of the Sample.* (See Table 1-6.)
- *Net Heating Value (NHV) = Lower Heating Value (LHV)* is the net heat released by a fuel compensating for the quantities of heat remaining in the vaporous state (as water vapor) in the flue gas.

Laboratory determination of LHV is difficult. When required, LHV may be calculated as follows:

$$Q_L = Q_H - 1040W$$

where:

Q_L = LHV of the fuel, Btu/lb
Q_H = HHV of the fuel, Btu/lb
W = lb water *formed* per lb of fuel combusted on stoichiometric basis
1040 = a factor commonly used to reduce high heat value at 80°F and constant volume to low heat value at constant pressure.[64]

For boiler combustion calculations, HHV is used as it is the basis on which fuels are bought and sold. LHV's are customarily used in Europe. When considering purchase of foreign designed and/or manufactured equipment, the buyer should maintain awareness of these conventions.

On an as delivered basis, the HHV's of wood fuel will range over the values listed in Table 1-6.

Note the sensitivity of HHV to moisture content. As we now can quantify, HHV is proportionally affected by fuel moisture content.

Comparisons of the HHV's of various fuel woods and their equivalents in tons of coal are shown in Table 1-7.

PHYSICAL/CHEMICAL CONSIDERATIONS

Combustion of wood fuel occurs in several stages (Figure 1-3). The surface of the wood initially undergoes thermal breakdown into vapors, gases, and mists, some of which are combustible. The first stage of combustion exists up to 395°F. In this zone, there is a slow, steady weight loss as water vapor and other non-ignitable gases are driven off.[73]

In the temperature range of 395 to 535°F more gases are driven off and heat liberating reactions first occur; however, there is no flaming until higher temperatures are reached. In the next zone, temperatures range from 535 to 935°F; in this zone, gases continue to evolve and react, giving heat. At first, they are too rich in carbon dioxide and water vapor to sustain flame, but secondary reactions occur, forming combus-

TABLE 1-6. Available Energy in a Wood Fuel at Different Moisture Contents.[8]

M.C. WET BASIS (%)	HIGHER HEATING VALUE (HHV) (BTU/LB)	M.C. DRY BASIS (%)
0	8,750	0
20	7,000	25
50	4,375	100
80	1,750	400

TABLE 1-7. Approximate Weights and Heating Values Per Cord of Fuel Woods.[49]

VARIETY OF WOOD	WEIGHT PER CORD, LB		AVAILABLE HEAT UNITS PER CORD, MILLION BTU (GHV)		EQUIVALENT IN HEAT VALUE TO TONS OF COAL*	
	GREEN	AIR DRY	GREEN	AIR DRY	GREEN	AIR DRY
Ash, white	4300	3800	19.0	20.5	0.77	0.79
Beech	5000	3900	19.7	20.9	.76	.80
Birch, yellow	5100	4000	19.4	20.9	.75	.80
Chestnut	4900	2700	12.9	15.6	.50	.60
Cottonwood	4200	2500	12.7	15.0	.49	.58
Elm, white	4400	3100	15.8	17.7	.61	.68
Hickory	5700	4600	23.1	24.8	.89	.95
Maple, sugar	5000	3900	20.4	21.8	.78	.84
Maple, red	4700	3200	17.6	19.1	.68	.73
Oak, red	5800	3900	19.6	21.7	.75	.83
Oak, white	5600	4300	22.4	23.9	.86	.92
Pine, yellow	—	—	21.1	22.0	.81	.85
Pine, white	—	—	12.9	14.2	.50	.55
Walnut, black	—	—	18.6	20.8	.72	.80
Willow	4600	2300	10.9	13.5	.42	.52

*Short ton (2,000 lb) of coal having a heating value of 13,000 Btu/lb.

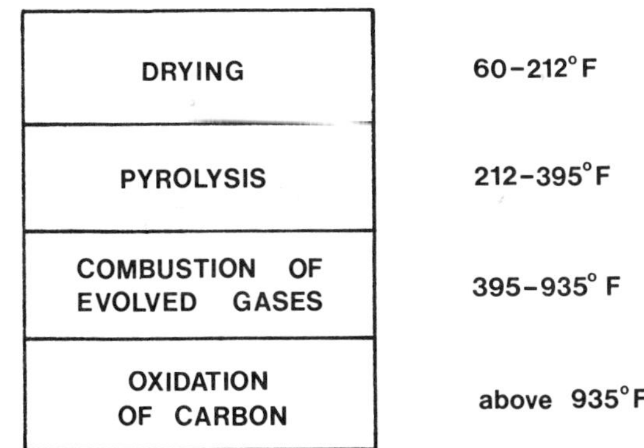

Figure 1-3. Stages of wood fuel combustion.

tible gases that ignite and flame. Finally, all the gases and tars are driven from the wood, and pure carbon (usually referred to as charcoal) remains. Combustion of the charcoal occurs, and the temperature of the wood surface rises above 935°F. Since the four stages of combustion occur simultaneously, many secondary reactions result which further complicate combustion. Though all direct combustion of wood occurs in these four stages, there are sufficient differences between available wood fuels and heating applications to require different types of combustion equipment (see chapter 2).

2

Wood Combustion Equipment for Steam and Process Heat

Equipment discussed in this chapter includes:

- Wood-fired package boilers
- Equipment for retrofit of existing coal boilers to wood fuel
- Cyclone and suspension burners
- Wood gasifiers and pyrolysis units
- Fluidized bed combustors
- Other combustion systems

For this discussion, applications of the above are for plants having steam requirements up to 50,000 lbs/hr. Larger wood-fueled power plants generally mandate field-erected boilers. Field-erected units—common in the pulp and paper industry fall in a separate economic category, outside the scope of this book.

BOILER TYPES

Firetube and Watertube

Boilers may be divided into two general classes—firetube boilers and watertube boilers. In the firetube design, the heated gases travel through steel tubes passing through a water jacket, while in the watertube design the water being heated passes through steel tubes heated on the outside by the hot gases from the combustion process (Figure 2-1). There are many applications where the firetube unit has distinct advantages over a

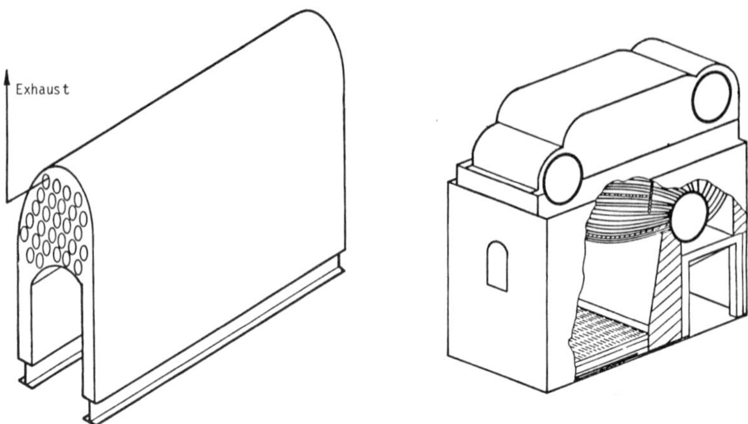

Figure 2-1. Firetube and water tube boilers.

watertube unit, particularly in light and medium industrial environments. The firetube boiler can be cheaper to purchase initially and it can be cheaper in terms of routine maintenance, particularly with regard to water treatment. Its limitations become apparent in the 20,000 to 40,000 lb/hr range, at pressures exceeding 300 psi. Larger shell diameters require thicker plates to withstand pressure and temperature stresses. Temperature differentials in the boiler create high stresses, and these stresses combined with the effects of precipitates and other deposits, have caused boiler explosions in the past. Because of its smaller component sizes and ability to accommodate expansion, the steel watertube boiler is more suitable for large capacities and high pressure.

Package and Field-Erected Units

Besides the firetube and watertube classifications, boiler designations can be made in another manner: boilers can be *package* boilers or *field-erected* boilers. These designations cause some confusion since virtually all wood-burning units require some field erection. A package boiler generally can be shipped by normal transportation methods such as a flat-bed truck or railcar. The major boiler components are in one assembly and can be lifted onto a simple foundation and piped into an existing system. As a result, the package boiler requires less labor before start-up than a field-erected unit. The field-erected unit often requires individual welding of boiler tubes to the tube sheets and the entire fabrication of a steel framework. The component parts of a field-erected boiler are completely built up at the job site, while the package boiler is nearly complete when it leaves the factory. Package boilers in the 200,000 lb/hr range have been shipped for gas/oil firing, but the larger combustion volumes necessary for wood units generally limit the size for a wood-fired package boiler to less than 50,000 lb/hr. As one might expect, field-erected boilers cost more ($/lb-steam) than package boilers, and construction times are significantly longer.

Burners and Grates

Finally, boilers may be differentiated by the method of combustion. Typical designs are pile burners, cyclone and suspension burners, and fluidized-bed combustors. Each system is optimally suited to particular variations of fuel inputs and types of process heat output desired.

Boilers incorporating a *pile burning design* (Figure 2-2) find applications where the anticipated wood fuel has a high (up to 65%) moisture content, as is found in whole green tree chips, bark, and green mill residues. Size control of the fuel is not so critical as in a cyclone or suspension burner.

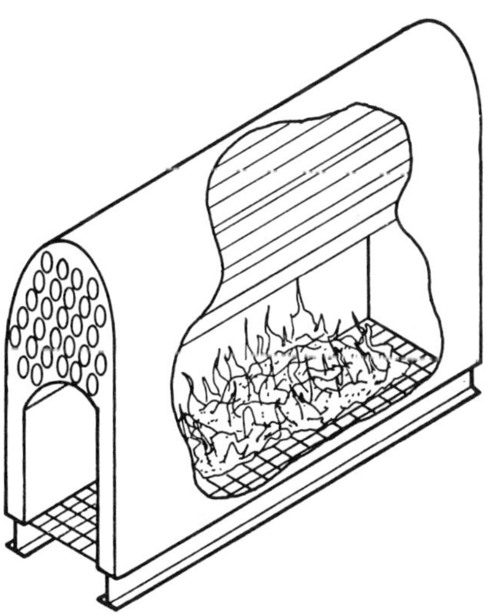

Figure 2-2. Pile burning on grate.

Grates serve to support the fuel while it dries and volatiles are driven off. Air flowing up through the grates (*underfire air*) serves several purposes, as follows:

- provides oxygen for combustion
- cools the grates
- promotes turbulence in the fuel bed, and
- contributes to drying the fuel.

The introduction of *overfire air* above the grate induces turbulence and provides oxygen for combustion of entrained and volatized material.

One type of heaped pile burner is the *dutch oven*. Usually fuel is gravity fed through a fuel chute onto the pile. Within the refractory-lined chamber, high temperatures are generated and the fuel is dried. Underfire air serves to partially burn the fuel and drive off the volatiles. Burning is completed in a secondary chamber where overfire air is injected (Figure 2-3).[38]

The major advantage of a dutch oven is its ability to utilize wet fuel of a rough, chunky consistency. A problem with dutch ovens is their slow response to load swings—a result of the thermal *inertia* of the fuel pile. The fuel/air ratio changes as the fuel pile burns down, thus making control difficult. The most serious drawback to the dutch oven is low

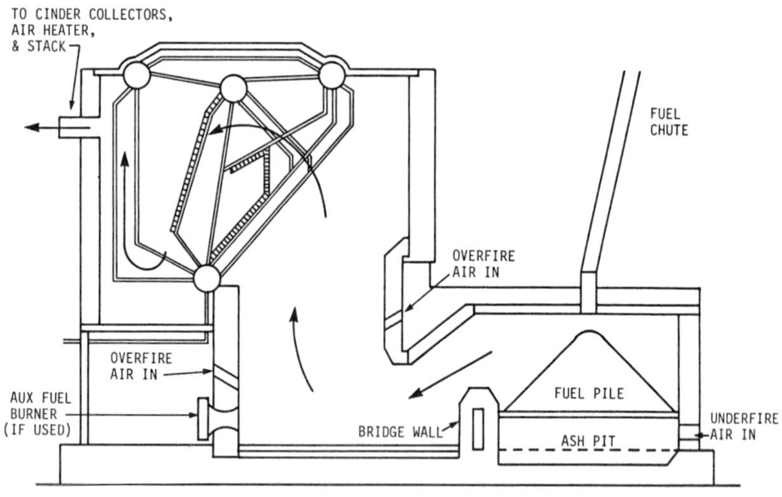

Figure 2-3. Dutch oven.

efficiency (50 to 60%). The low efficiency results from: (1) increased heat loss due to the furnace's large surface area and (2) absence of radiant heating (the furnace and boiler are separate). The pile burning combustor combines well with a firetube design for steam generation, both in terms of moderate first cost and economy of size. *Turndown ratios* (the minimum percentage of the maximum design load at which a boiler will operate efficiently) may be limited, and the ability of the unit to accept coal should not be taken for granted. A dutch oven unit is illustrated in Figure 2-3.

Gravity feed of fuel onto the pile causes two pollution problems: (1) increased probability of unburned particles being entrained and leaving the combustion zone as particulates and (2) cooling of the combustion zone and hindrance of complete combustion. These problems can be effectively eliminated by pushing the fuel onto the pile from underneath. This is accomplished through the use of an underfed stoker.

The second class of pile burners has a thinly spread pile. Typical thin pile burners would be dutch oven or field-erected boilers with sloping grates (Figure 2-4). Fuel slides into the furnace on the grate. Particulate problems associated with dropping fuel in from above are eliminated. The slope of the grate is a function of the fuel condition. Since dry fuel slips easier than wet fuel, the grate is designed with different slopes in the drying and combustion zones. The thin pile allows more uniform air distribution as compared to a heaped pile, and combustion rates can be increased more rapidly. Wet fuel can be used, but more size uniformity is required than with a gravity-fed dutch oven. However, a problem exists with preventing blowholes in the fuel bed, especially with non-uniform fuel. Another problem is boiler size limitations since sloping grate boilers have not been made in sizes as large as spreader stoker boilers.

Most modern boilers (in the 20,000 to 50,000 lb/hr range) combusting coal use some type of moving grate design. Variations include *traveling grates, dump grates,* and *rotary grates* (Figure 2-5). The majority of large wood-fired boilers utilize *spreader stokers,* however. The stoker can be mechanical or pneumatic. Mechanical spreader stokers resemble a paddle wheel and ''throw'' fuel into the boiler, while pneumatic spreader stokers use air pressure to ''blow'' fuel into the boiler. Pneumatic stokers find wider application with wood fuel due to the size inconsistency of wood. These boilers have a high heat release rate because the smaller particles burn in suspension. The heavier particles fall to a

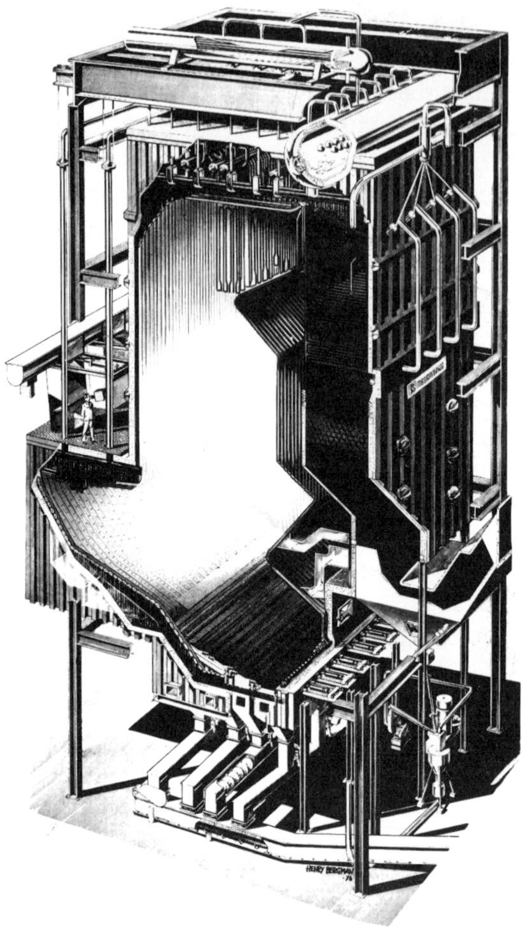

Figure 2-4. Sloping grate bark boiler. (*courtesy Maskinverken Co.*)

grate where they are burned in a thin bed. Spreader stoker installations can burn wet fuel, with some drying achieved with pre-heated combustion air. High heat release rates, integral furnace and boiler, and lack of refractory all contribute to a smaller and lighter boiler. In large-scale boilers (greater than 80,000 lb-steam/hr), spreader stokers are used in conjunction with traveling grates to simplify ash removal.

Disadvantages of spreader stokers are as follow: (1) fuel interruption can extinguish the flame in the absence of refractory; (2) overfeed leads to heavier flue gas particulate loading; (3) heavier hogged fuel particles

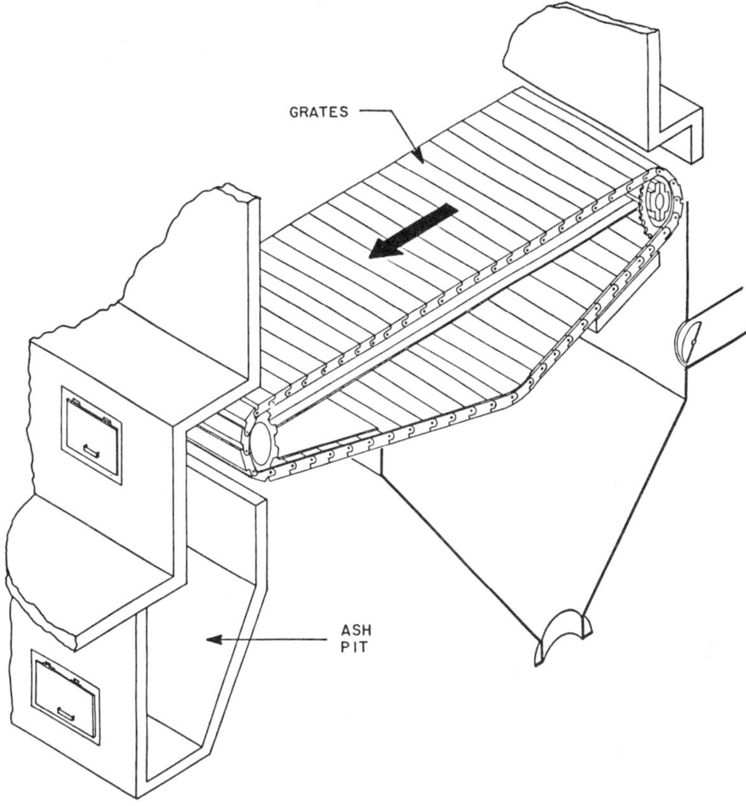

Figure 2-5. Traveling grate, schematic.

can fall straight down and be partially unburned when dumped since traveling grates generally rotate back-to-front, and (4) traveling grates have high maintenance costs.

Moving grates are generally installed on watertube steam generators; these units respond quickly to load changes and have high turn-down ratios. With minor modifications to the fuel feed systems, units designed for coal will work well with properly sized wood fuels. Should a high moisture wood fuel (50% M.C.) such as whole green tree chips be substituted for coal, the boiler will inevitably undergo derating, and steam output may be reduced by as much as 30% to 40%. If a dry wood fuel is used, it should be possible to make 100% of boiler rating. Plant engineers have reported good success mixing dry (15% M.C.) wood pellets 50/50 with coal. Under such conditions, full boiler ratings were made and, most important, reduced stack emissions of both particulates and SO_x were achieved. Wood pellets have a higher cost ($/MMBtu) than coal; however, it has been reported that operating on the 50/50 coal/wood pellet mixture, additional pollution control devices such as baghouses (for particulate) and FGD units (flue gas desulfurization) become unnecessary.

Cyclone and suspension burners are first cousins to pulverized coal burners. Dry (less than 15% M.C.), properly sized (less than ¼″) wood fuel is turbulently mixed with forced air and either combusted in a stream over the main fuel bed (suspension burners) or mixed in the first stage of the burner and combusted in an external cyclone burner. Cyclone burners have been successfully applied in providing heat for brick kilns in Georgia and South Carolina. Small particle-sized, dry wood fuel has high explosion potential; therefore, a flame safeguard device (such as ''Fireye'' by AC controls) must be provided. Boilers fired with cyclone and/or suspension burners often have high turndown ratios, good efficiencies, and excellent response to swing loads.

Suspension burners suffer the same problems as the overhead feed pile burners. Entrainment of particulates makes control of fly ash diffi-

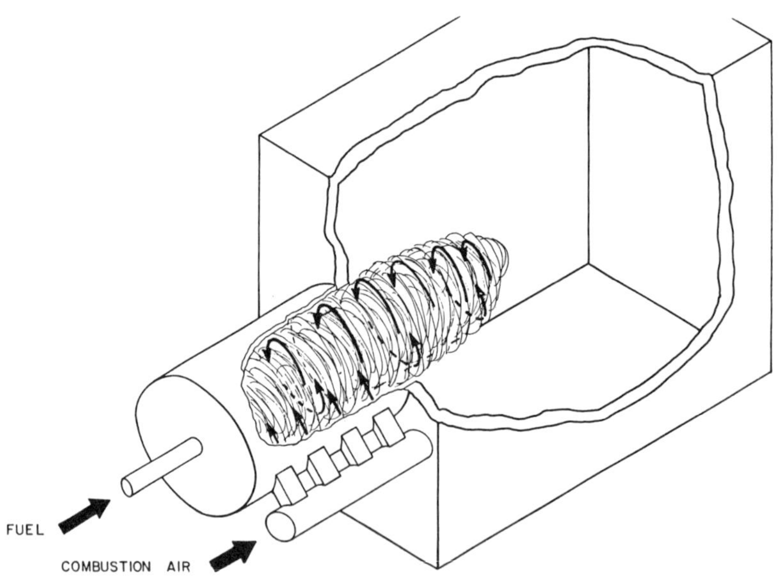

Figure 2-6. Cyclone burner, schematic.

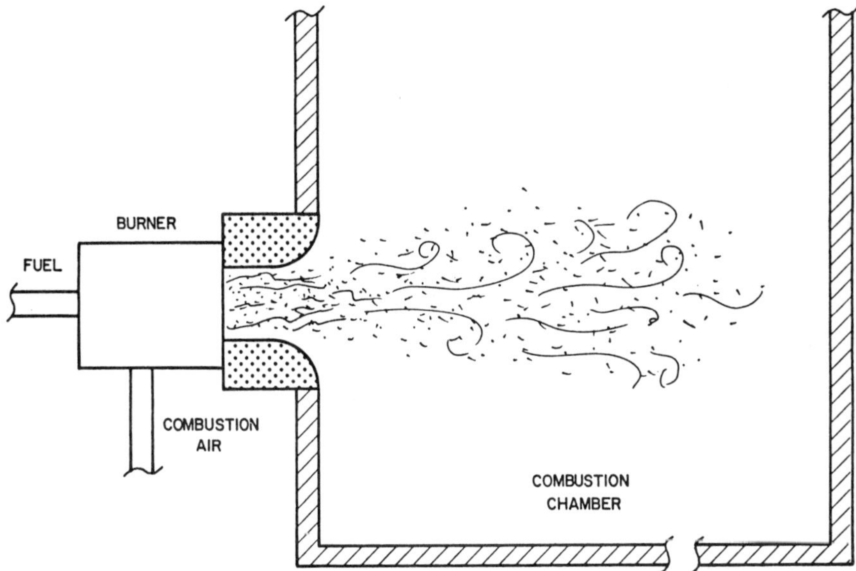

Figure 2-7. Suspension burner.

cult; and, at high combustion rates, the residence time of fuel may be insufficient for complete combustion. Cyclone and suspension burners are illustrated in Figures 2-6 and 2-7.

WOOD-FIRED PACKAGE BOILERS-MANUFACTURERS

Wood-fired package boilers represent proven, commercially available, reliable technology. Many vendors are active in the wood package boiler field. In general, the larger and perhaps better known boiler manufacturers (Babcock and Wilcox, Combustion Engineering, Riley Stoker, Foster-Wheeler, etc.) are not really interested in building small (less than 50,000 lb/hr) boilers for light- and medium-sized commercial and industrial operations. This gap has been filled by a number of smaller manufacturers, and the market has become quite competitive. Depending on the type of wood fuel to be used, an industrial customer should be able to receive at least several bids on a complete wood burning system. The cost will be substantially higher than the traditional gas/oil package boiler, but the payback time due to fuel cost savings can be attractive.

Most wood-fired boilers on the market are automated or nearly automated. A full-time boiler operator is usually required, however. System malfunctions which may occur will often be in the wood handling system. Most package boilers will require flyash collection devices to meet local air pollution codes, and the prospective industrial customer should investigate all local and state regulations before signing a contract. In addition, the contract should ensure that the boiler manufacturer or builder will guarantee compliance with the air pollution regulations.

The first cost of wood burning package boilers can be quite high. The less expensive range of boilers are generally the ones with the least flexibility with regard to the quality (i.e., moisture content, size, etc.) of fuel that may be burned. As with suspension wood burners, even the smallest package boilers require a given amount of solids handling equipment and control systems. The system cost per pound of steam is much higher for the smaller systems. A given size of wood package boiler will generally have a first cost of four to five times that of a comparable gas/oil boiler.

It may be difficult to retrofit an existing steam plant with a wood fired package boiler due to space limitations. Wood systems generally require a large area for wood handling and storage, and the combustion volumes are much greater than they are for gas or oil, dictating a larger physical boiler/combustion chamber size for a given steam output.

(The manufacturers and equipment listed below are presented as a representative sample for illustrative purposes only. It is by no means a comprehensive survey of available equipment. No endorsement of any particular unit, express or implied, is meant.)

Industrial Boiler Company

The Industrial Boiler Company of Thomasville, Georgia, has installed over 65 wood-fired boiler systems in 23 states over the last several years. The company's primary product for the wood industry is an HRT type firetube boiler (Figure 2-8) which is available in sizes from approximately 2,600 to 34,000 lb steam/hr in single boiler installations. The main boiler components are factory-made and field-erected. Different options are available for grate designs and furnaces. Some installations use dutch ovens which require more refractory work. Operating pressures range up to 300 psi, and most installations have been able to meet local pollution codes with strictly mechanical pollution control equip-

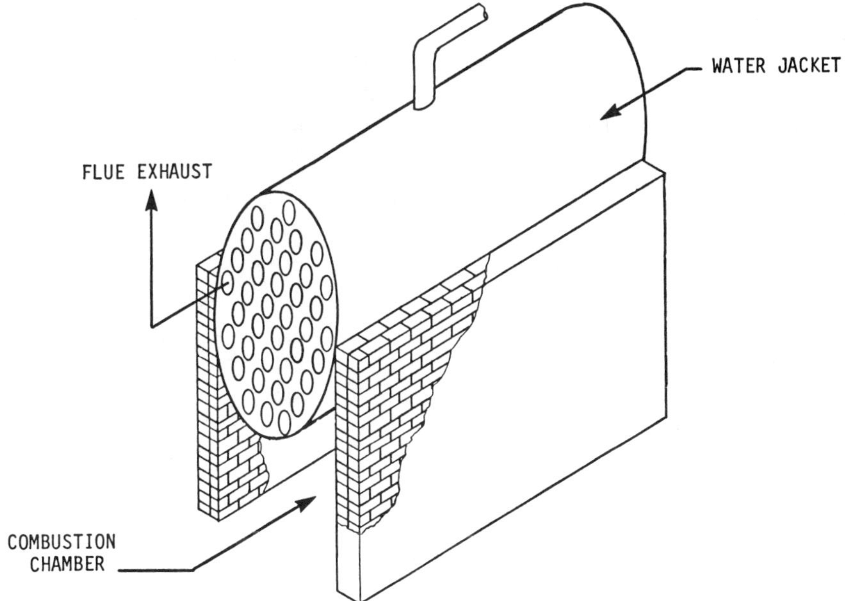

Figure 2-8. Horizontal return tubular boiler.

ment. The company will install turnkey jobs which include complete wood handling and storage systems, controls, and air pollution equipment. The manufacturer claims a thermal efficiency of 60%, and these units can burn wood waste up to a moisture content of 60%.

Advantages of the Industrial Boiler units include simplicity of design and associated low maintenance costs. Replacement of boiler tubes is a straightforward operation, and the largest single maintenance items probably will involve refractory repairs. The smaller Industrial Boiler units are semiautomatic and require little operator attention. A representative Industrial Boiler Co. installation is shown in Figure 2-9.

Ray Burner Company

The Ray Burner Company of San Francisco, California, offers a packaged boiler for wood use which is similar to many gas/oil fired boilers in appearance. The Ray unit is a three pass dry-back firetube boiler available in sizes from 100 to 500 hp (3,400 to 17,000 lb steam/hr). Wood waste is dropped into the combustion chamber from above and

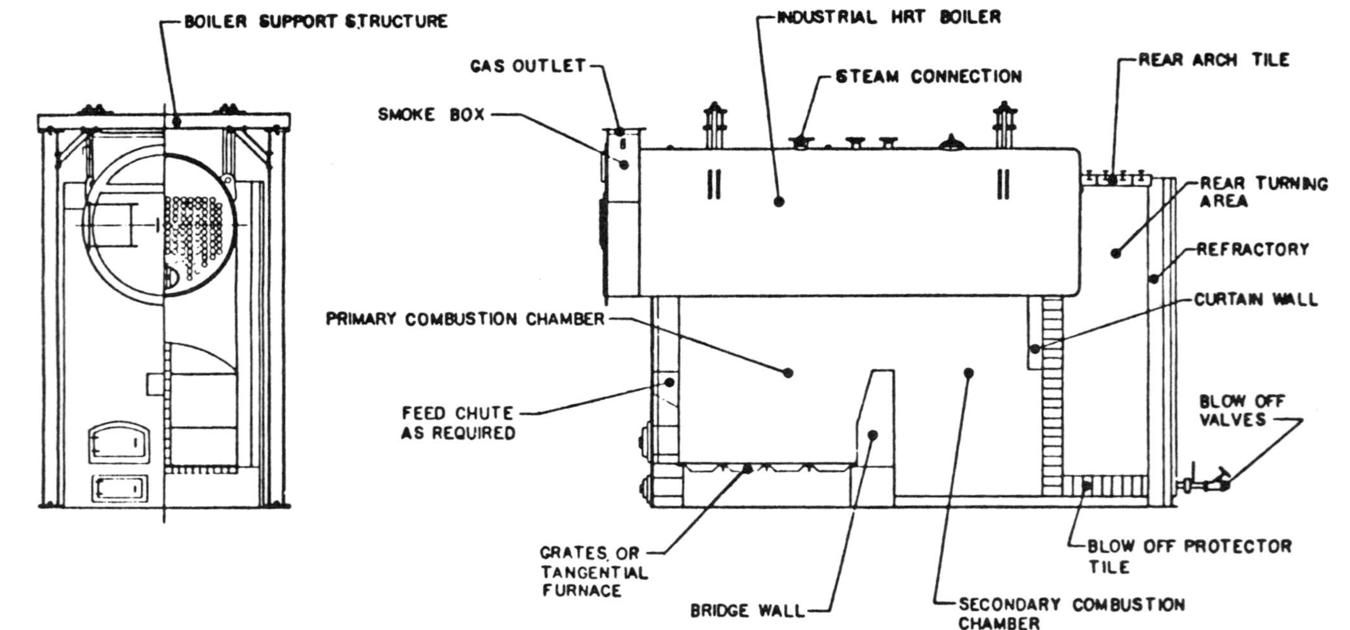

Figure 2-9. HRT Boiler for wood firing. (*courtesy Industrial Boiler Co.*)

burns partially in suspension and partially on the stationary grate. Auxiliary oil/gas firing is available and is used for startup or instances of wood fuel interruption.

The fuel must be 20% moisture content or less, which makes this unit ideal for furniture or finished lumber plants. (Predrying of the fuel would be necessary for the combustion of green chips or wet wood waste). The manufacturer claims a thermal efficiency of 80% (quite believable when dry wood is burned) and also claims that additional cleanup equipment (other than the cyclone system supplied with the boiler) is rarely required to meet air pollution codes. The manufacturer lists more than 25 successful installations to date.

An advantage of the Ray Burner units is the total package design which allows the easy installation of the unit in place of a gas/oil boiler. A cross-section of a typical Ray Burner wood boiler is shown in Figure 2-10.

The Bethelehem Corporation

The Boiler Division of the Bethlehem Corporation in Easton, Pennsylvania, markets a package boiler for burning wood waste based on coal burning technology developed in England by G.W.B. Ltd. in the 1950s. The Bethlehem boiler is a three pass firetube boiler of a ''wetback'' configuration. The wetback design is reported to give greater flexibility to the pressure parts, which in turn lowers maintenance requirements and eliminates part of the refractory.

Fuel is introduced into the boiler through the top of the combustion chamber with a pneumatic conveying system which spreads the fuel over the burning bed. Integral with the boiler, a mechanical flue gas cleaner is provided which allows for recirculation of the collected flyash to the combustion chamber. Ash removal from the combustion chamber is accomplished manually. When wood is being burned, the manufacturer claims ash removal can be done once per shift.

Bethlehem supplies multifuel boilers in sizes up to 30,000 lb/hr of steam. A turndown ratio of 3/1 is claimed for wood fuel. As the boilers have a thermal efficiency of up to 83% on coal, they can be expected to be slightly less efficient on wood, dependent on the moisture content. The manufacturer claims that wood waste with a moisture content of up to 25% can be burned in the boiler.

A typical Bethlehem installation is included in Figure 2-11.

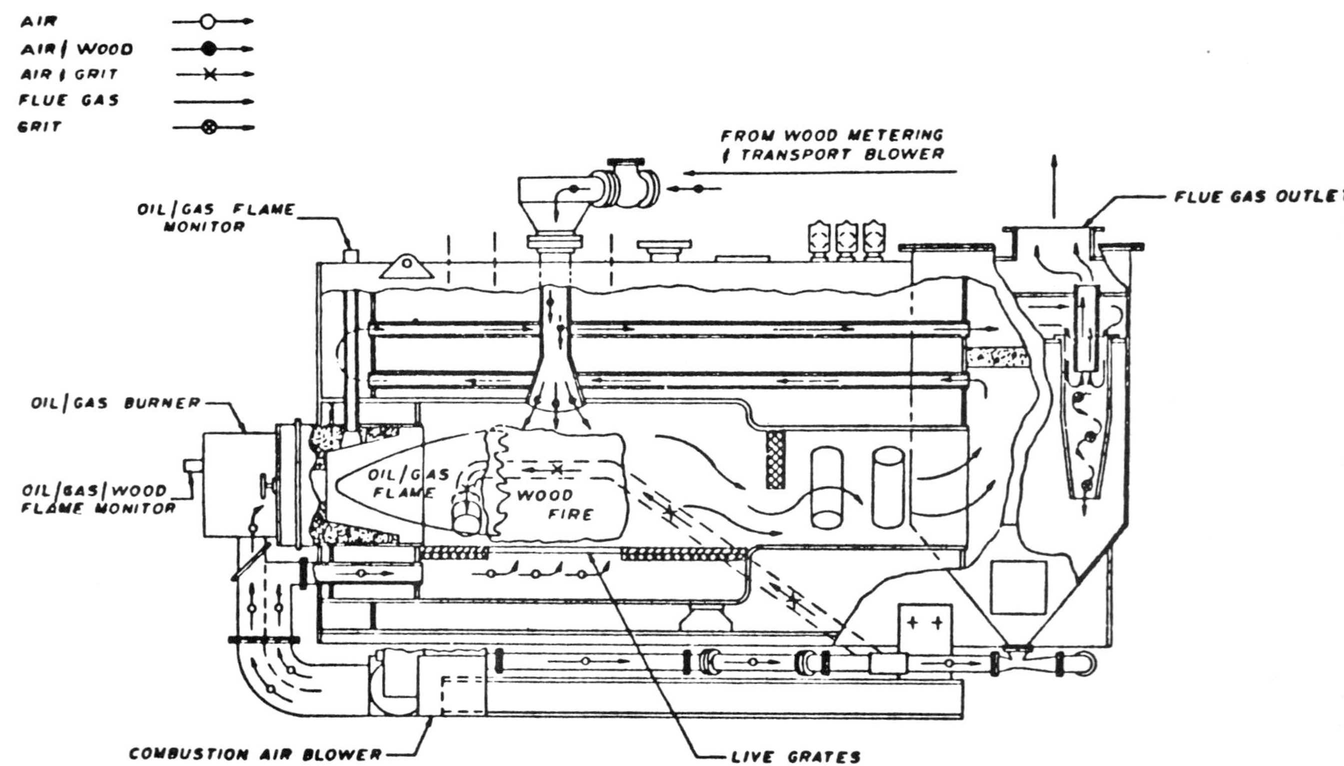

Figure 2-10. Firetube wood boiler. (*courtesy of Ray Burner Co.*)

Figure 2-11. Bethlehem wood waste boiler. (*with permission of the Bethlehem Corporation, Easton, Pa.*)

CNB Tri-Fuel Boiler

The CNB Tri-Fuel Boiler is manufactured by the Combustion Service & Equipment Company in Pittsburgh, Pennsylvania. It was originally marketed as a boiler able to fire coal as well as oil and gas, but several successful installations have been made using wood as a solid fuel. The boiler is a firetube firebox of three-pass wetback design and is available in sizes from 3,400 to 23,000 lb steam/hr and in a high pressure (up to 200 psi) or low pressure (15 psi or hot water) configuration.

The CNB boiler is equipped with an underfeed stoker for handling wood waste, wood pellets, or coal. A gas/oil integral fan burner is also provided for standby and load swing control.

The manufacturer claims a thermal efficiency of 80% for this boiler firing wood waste. Air pollution regulations are met with no additional collectors, and the boiler is reportedly capable of burning 100% wood waste with a moisture content of 30%. Wetter material can be burned if

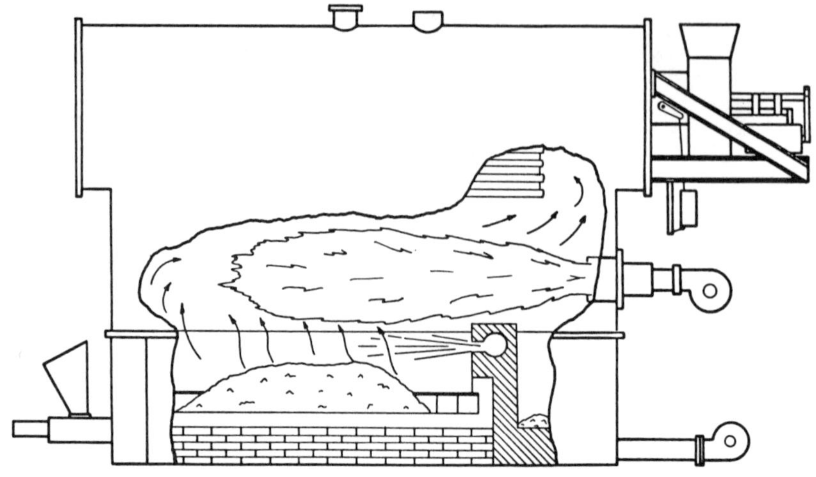

Figure 2-12. CNB Tri-fuel boiler.

auxiliary fuel is co-fired. A typical Tri-Fuel boiler illustrating the overall design is shown in Figure 2-12.

Wellons Boilers

Another approach to the wood burning boiler concept is taken by Wellons, Inc., of Sherwood, Oregon. At present, more than 130 wood boilers have been started up by this company, and it is also well known for its lumber dry kilns and wood storage bins. The size range available for their boilers is 3,000 to 60,000 lb/hr of steam, with larger systems possible using multiple boilers.

Wellons uses a pile burning system company officials call the "cyclo-blast furnace," which they report is capable of burning wood up to 50% in moisture content with a maximum particle size of 3″. The furnace is a refractory chamber which allows complete combustion of the fuel, but reportedly limits turndown ratios (although the manufacturer claims 5/1). Load response, due to the pile burning design, is probably slower than with other boiler configurations.

The Wellons furnace is a watertube design and is equipped with combustion air preheating. Air pollution requirements are met with a multicyclone mechanical collector and the furnaces are operated at balanced draft with I.D. fan and F.D. fans. Figure 2-13 gives an overall view of a typical Wellons system.

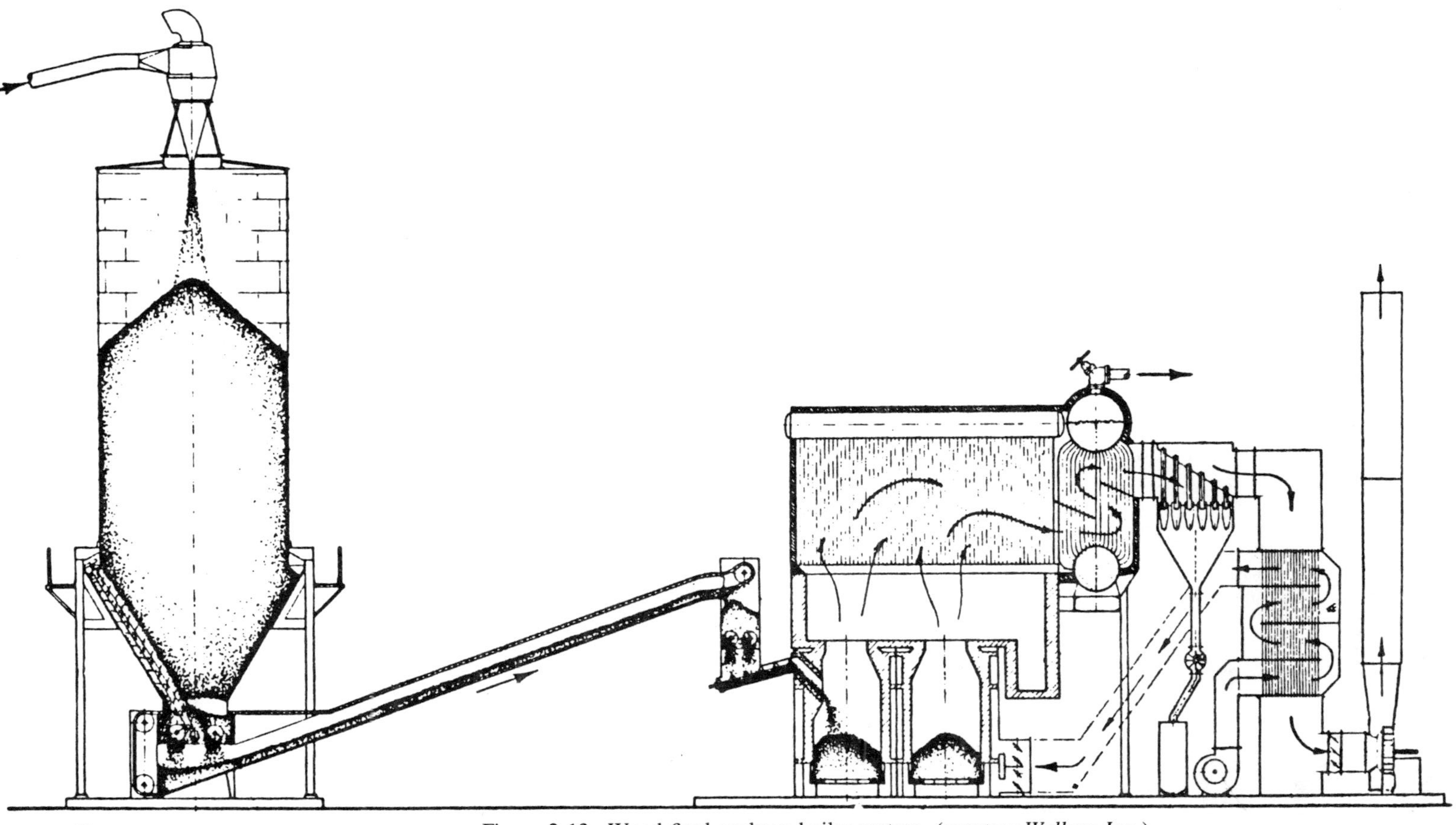

Figure 2-13. Wood fired package boiler system. (*courtesy Wellons Inc.*)

Deltak Corporation

The Deltak Corporation of Minneapolis, Minnesota, offers a packaged boiler system called the Charger. It is very adaptable and quite different, since Deltak has customarily dealt with the design problems involved with waste heat recovery in their waste heat boiler systems. The Charger features shop-assembled components, developed primarily for solid fuel combustion. Its operational range is 30,000 to 110,000 lb/hr of steam, with the ability to produce high pressure steam—ideal for cogeneration.

These boilers are designed for fuel to be fed by any of the conventional wood stokers. The boiler can be auxiliary fired with fossil fuel—thus allowing for wet wood fuel to be burned. Deltak claims that wood waste with a moisture content of up to 55% can be burned and that boiler efficiency is in the range of 65%.

A typical Deltak boiler installation is shown in Figure 2-14.

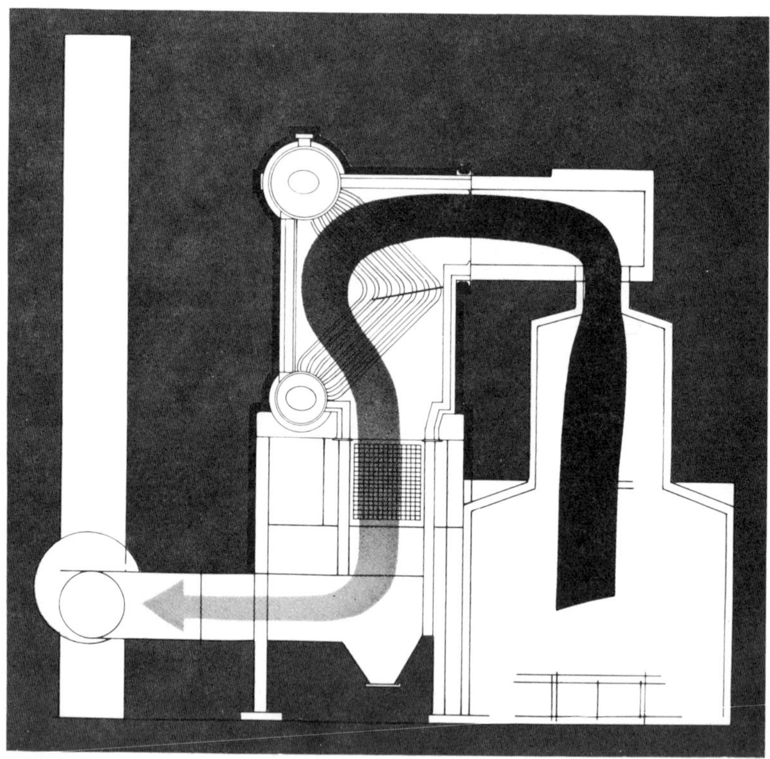

Figure 2-14. Modular wood fired package boiler. (*courtesy Deltak Corp.*)

Weiss Boiler Company

The Gebr. Weiss Company of Fronhausen, West Germany, has been supplying boilers to the woodworking industry in Europe for many years and has begun installation of its units in this country. Weiss manufactures a line of watertube boilers called the "Compact" series which are available in configurations ranging in size from 2,400 to 50,000 lb/hr. These installations are customized depending upon the customer's needs and are available with shaft furnaces, inclined grates, or step grates. Auxiliary firing of oil or gas is provided with most of these units. Depending on the fuel and the regulations, most of these units can meet particulate standards with mechanical collectors. A typical Weiss "Compact" installation is shown in Figure 2-15.

Figure 2-15. "Compact" wood waste boiler. (*courtesy Energy Control Engineering Corp.*)

Weiss also supplies large field-erected radiant boilers in capacities up to 140,000 lb steam/hr.

As with most large purchases of this type, the customer should visit a working installation similar to what theirs will become, if possible. Most boiler manufacturers will be glad to arrange such a visit. (A more complete list of boiler vendors is included in Appendix 1.)

SUSPENSION AND CYCLONE BURNERS

The cyclone furnace for burning pulverized coal has enjoyed widespread use on utility boilers for many years. The fuel is very finely ground and blown into the furnace almost as a gas, and, as a result, the combustion process is complete and efficient, and fly ash removal is easily dealt with.

Variations on the cyclone furnace concept have been developed for burning wood waste, but certain limitations can hamper the feasibility of using these systems in many applications. The wood residue must be dry (less than 15% moisture content on a wet basis), and it must be hammermilled or hogged to fairly fine particles. In spite of these requirements, several companies have placed a number of these units in industrial plants—largely in the forest products and brick (kilns) industries.

Energex. The Energex burner, developed jointly by the Moore-Oregon Corporation and Energex Ltd., has probably been installed in more plants than any other wood suspension burner. It consists of a single chamber cyclone and can burn dry (up to 15% M.C.) wood waste of ⅛ in. size or less. The wood residue is blown into the unit tangentially through a series of manifolds from a pneumatic conveying system. The burner itself is a refractory line chamber which must be preheated with auxiliary fuel before wood residue is introduced. Ash must be removed from the chamber by the operator periodically (every few days) but otherwise the operation is fairly automatic. Turndown ratios of five to one are reported, and discharge temperatures of the combustion gases range from 1,600 to 2,000°F, depending on fuel characteristics. These units have been successfully used for hog fuel dryers, veneer dryers, lumber dry kilns, and boilers. There is some carryover of ash, so if product contamination must be considered, the ash effects must be closely examined.

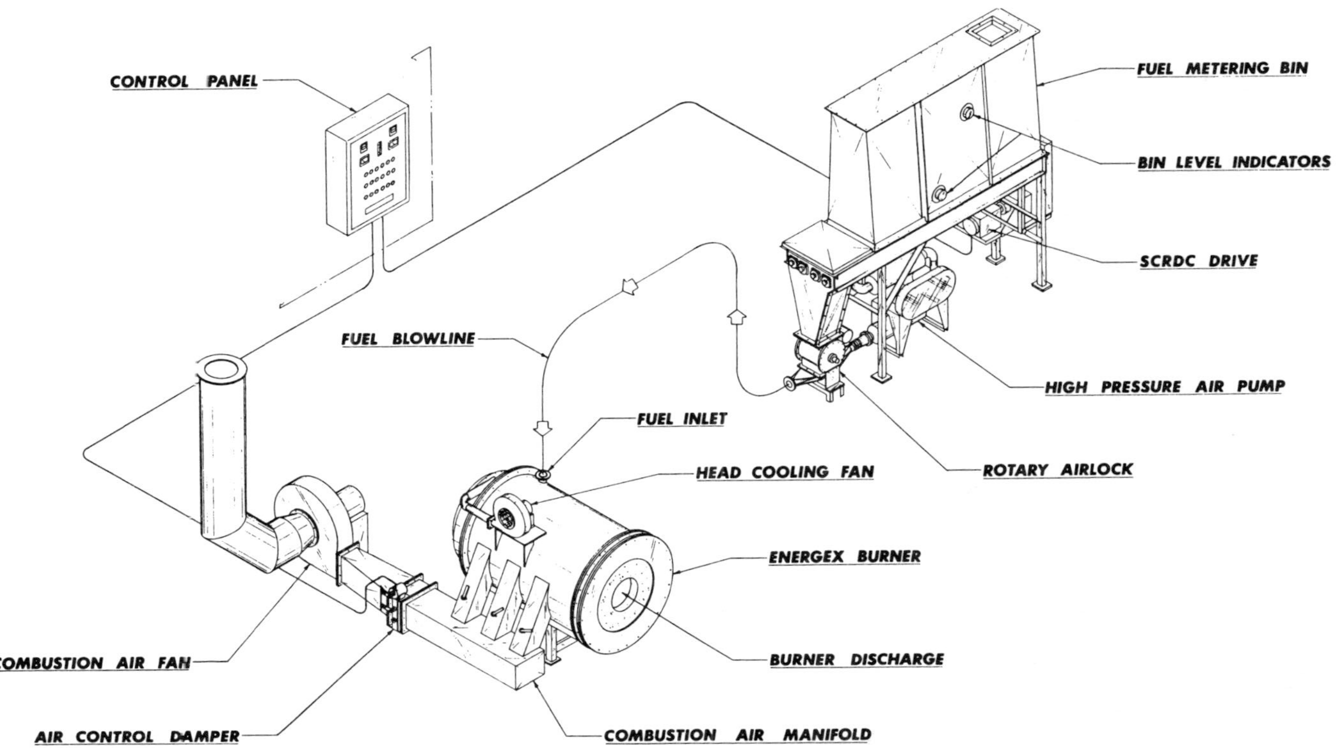

Figure 2-16. Energex burner. (*courtesy Moore International.*)

The Energex units are available in standard sizes of 6, 15, 27, 45, and 60 million Btu/hr. A schematic of a typical system is shown in Figure 2-16.

McConnell Industries. McConnell Industries in Birmingham, Alabama, has developed a cyclonic wood burner similar to the unit marketed by Energex. Present applications of these units are limited to the forest products industry where they are used to fire lumber dry kilns, veneer dryers, and rotary product dryers. The manufacturer indicates that performance is optimum when wood moisture content is 12% or less (wet basis). As with the Energex systems, the waste needs to be small in size. The manufacturer indicates that ash removal is necessary approximately once every week.

The McConnell Wood Burner has two refractory lined chambers. Ignition and nearly total combustion takes place in the primary chamber while the secondary chamber is used to complete the combustion. Four size ranges are available: 30 in. diameter (6–11 MMBtu/hr), 36 in. diameter (12–23 MMbtu/hr), 42 in. diameter (24–31 MMBtu/hr), and 48 in. diameter (32–40 MMBtu/hr). Plans are currently underway for producing larger sizes. An illustration of the McConnell system is shown in Figure 2-17.

Guaranty Performance. The Guaranty Performance Company of Independence, Kansas, markets a cyclone-type burner called the "ROEMMC" System which is similar in design to the units discussed above—with the exception of an integral ash separation device which reportedly results in a cleaner gas at the unit's output. Again, the fuel must be dry wood (less than 15% moisture content) with a uniformly small size. The manufacturer reports that the burner can run on a variety of materials including bark, peanut hulls, mesquite, peat, and several other categories of waste.

Due to the air pollution control equipment, the Guaranty unit is quite large and may be subject to higher heat losses than an Energex type unit. The burner system is available in sizes from less than 10 MMBtu/hr to greater than 60 MMBtu/hr. An overall view of a typical system is shown in Figure 2-18.

Coen Company. The Coen Company of Burlingame, California, is perhaps best known for its oil burning equipment and its low Btu gas

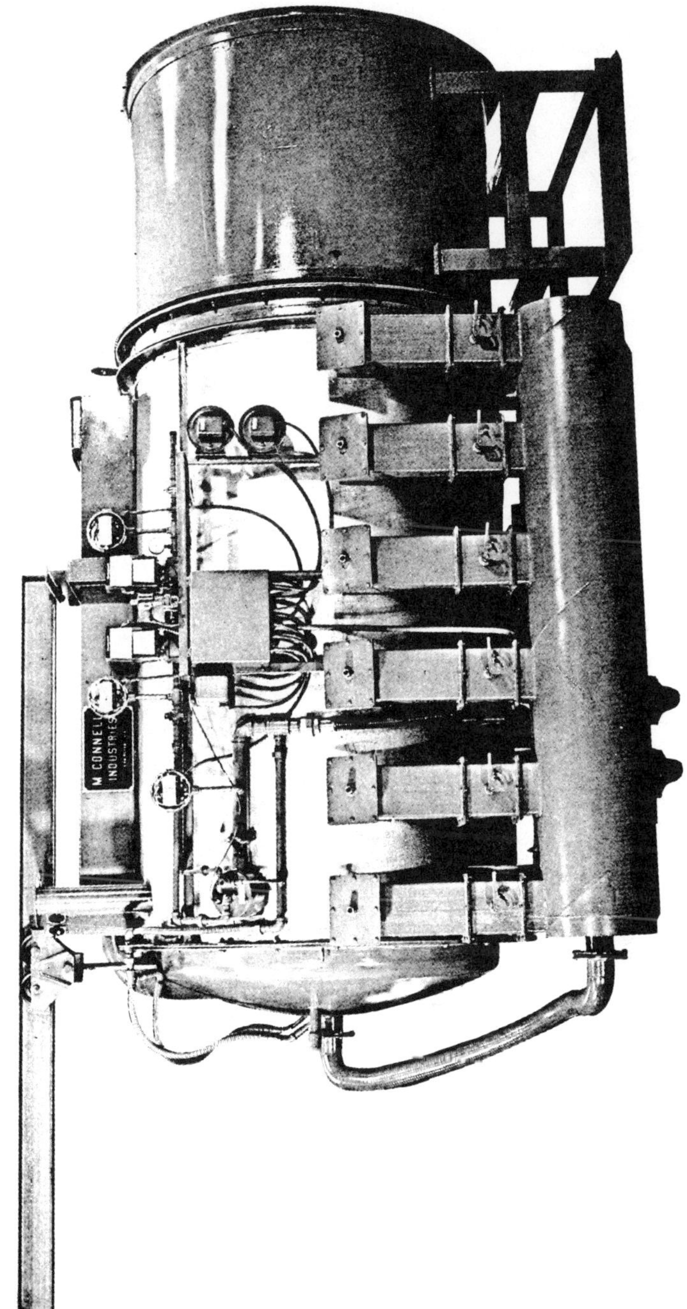

Figure 2-17. McConnell Wood Fuel Burner. (*designed and manufactured by McConnell Industries, Birmingham, Alabama.*)

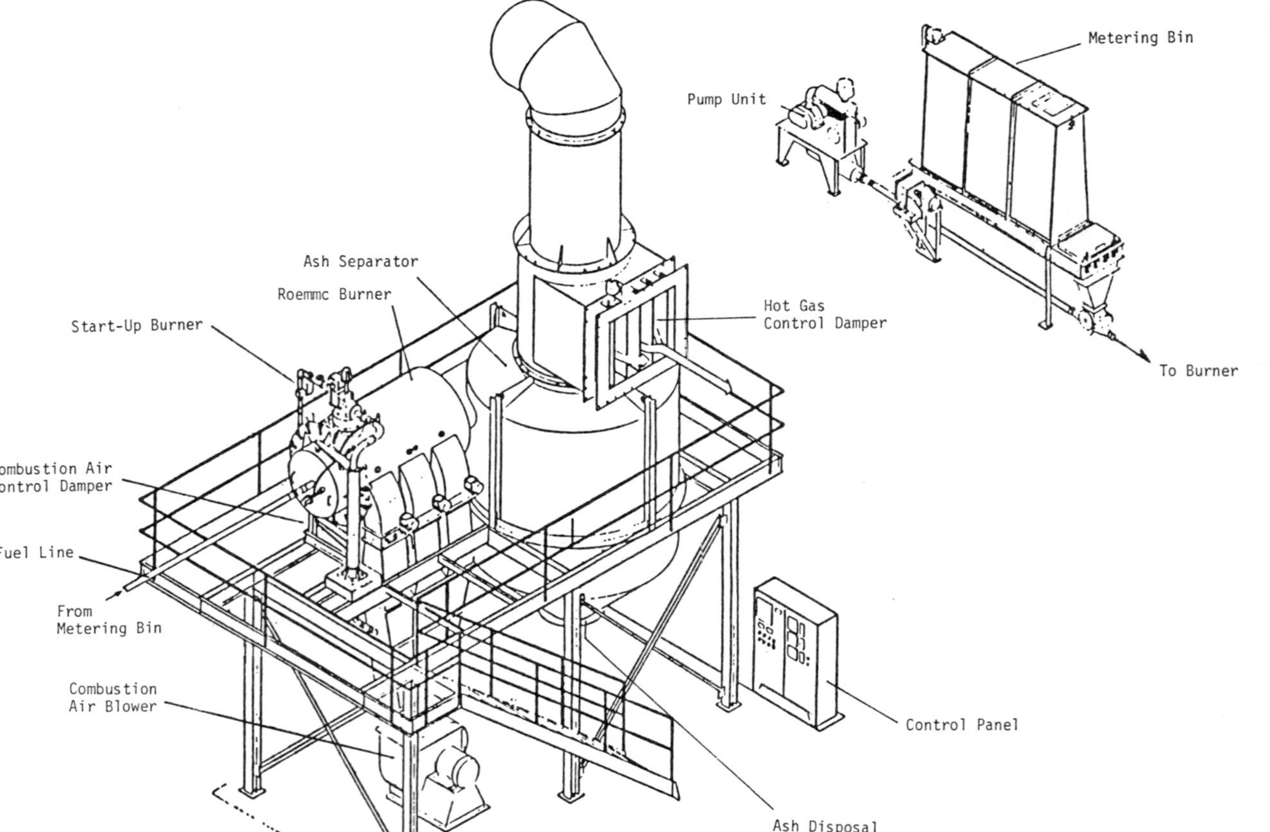

Figure 2-18. "ROEMMC" wood burner. (*courtesy Guaranty Performance Co.*)

burners, but it has been active in wood waste systems for some years and reports that it currently has more than sixty wood-burning systems installed.

The Coen burner is called the ''DAZ'' or ''Dual Air Zone'' and as with previous systems, it requires dry wood waste (less than 12% moisture content) for proper operation. The size of the particles that can be tolerated depends greatly on the application of the system, as larger particles can be burned where longer residence times are available. The Coen burner consists of two air registers with concentric louvers which divide the incoming air stream into two counterrotating concentric streams. These streams provide a turbulent mixing action which results in a compact flame pattern.

Coen reports that the DAZ burner has been fitted to incinerators, rotary dryers (for wood chips), veneer dryers, lumber dry kilns, field-erected boilers, and package boilers. Many installations include a refractory furnace or air heater which ensures complete combustion of the fuel. As with the systems already mentioned, periodic cleanout of ash from the combustion chambers will be necessary. Particulate emissions from the Coen units are highly dependent on fuel ash content. Coen reports that it can supply burners capable of providing 5 MMBtu/hr up to 100 MMBtu/hr, and will do turnkey jobs as well as supplying only the burner. A typical Coen installation is shown in Figure 2-19.

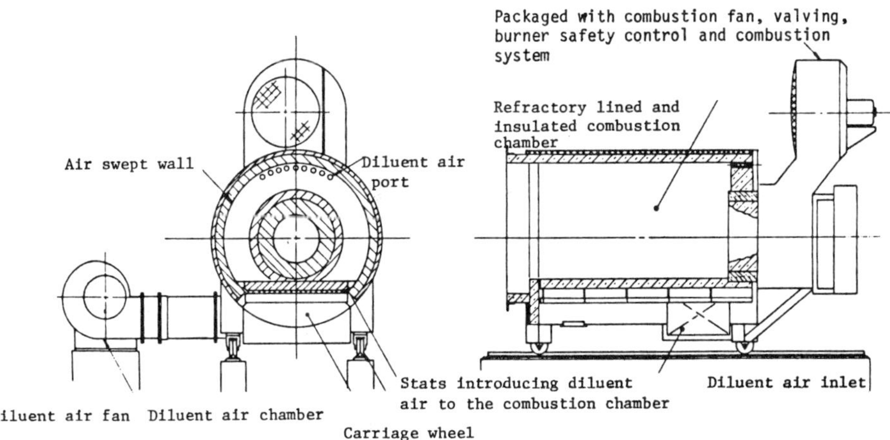

Figure 2-19. Dry fuel burner. (*courtesy Coen Co.*)

Peabody Gordon-Piatt. The Peabody Gordon-Piatt Company in Winfield, Kansas, manufactures a suspension burner with the capability of burning wood waste in conjunction with fuel oil or natural gas. Most of the installations of these units have been in boiler plants, although it appears that the systems could easily be adapted to veneer dryers or lumber dry kilns. As with the other systems in this section, the Peabody unit requires dry fuel (12% or less moisture content) with a maximum size of ½ to ⅛ in.

In the Peabody burner, the mixture of combustion air and wood fuel enters the burner fuel inlet under pressure. The scroll-type burner provides mixing of the fuel and air. This mixture is then propelled into the hot refractory zone of the burner where combustion is completed. Support fuel is used for startup and may be turned off (if proper wood fuel is available) afterwards.

Peabody Gordon-Piatt provides burners in the size ranges of 4.8 MMBtu/hr to 46 MMBtu/hr. For larger installations, multiple burners can be used. A number of control options are available and turndown ratios of 5/1 have been reported.

Figure 2-20 shows a layout of a typical Peabody Gordon-Piatt installation.

The above systems give an indication of what is available for burning dry wood waste. However, other systems are available, and many forest products plants build much of their own equipment for handling dry wood waste (or have it custom built). Prices for equipment vary widely due to the site specifics of the installation. The smaller systems have a higher cost per energy unit since they require roughly the same amounts of control and wood handling as the larger units.

In summary, the suspension and cyclone burners can be a viable alternative to the plant which already has a ready supply of dry wood waste available which could be reduced to a small size with a minimum of effort. Other manufacturers of this type of equipment are included in Appendix 1.

FLUIDIZED-BED COMBUSTORS

The fluidized-bed combustor (Figure 2-21) has enjoyed much publicity over the past several years, particularly as an alternate combustion system for coal burning. Basically, a fluidized bed relies on a combustion

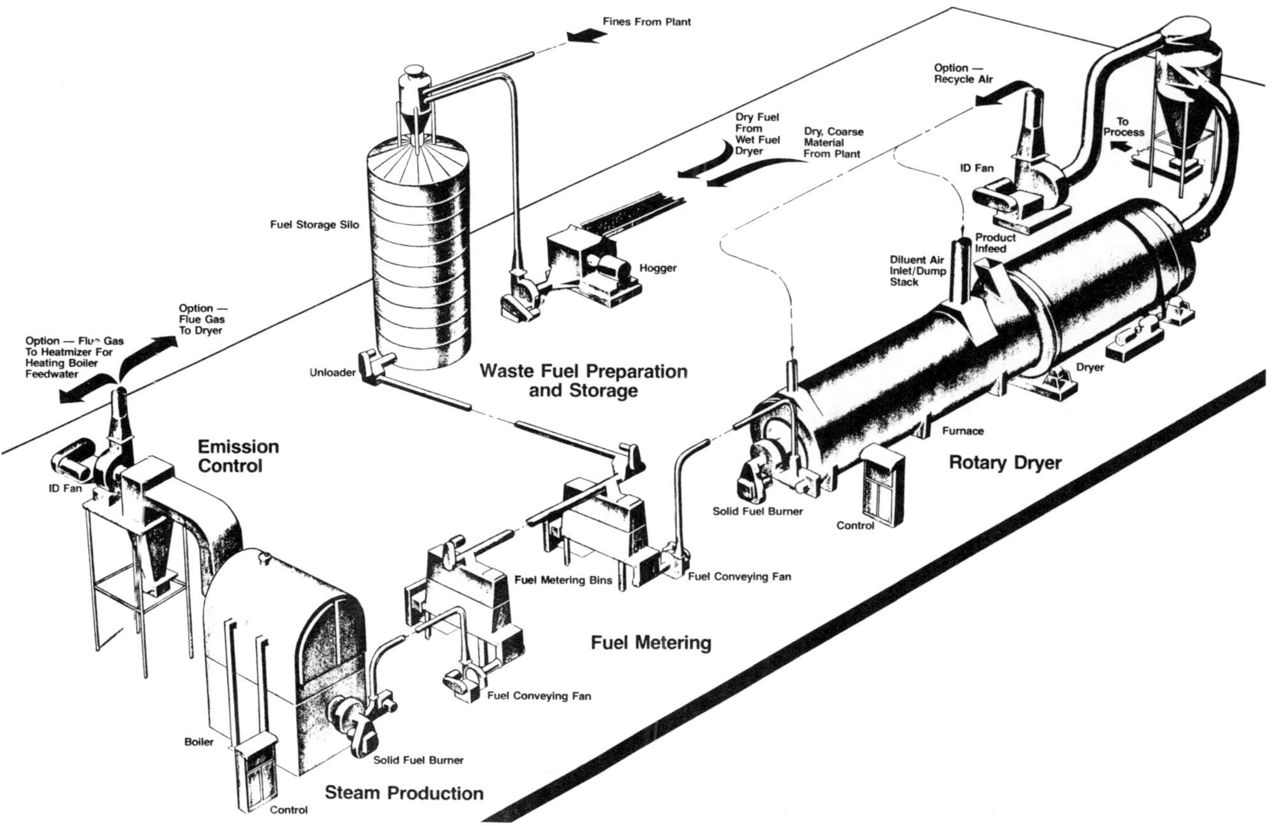

Figure 2-20. Dry wood combustion system. (*courtesy Peabody Gordon-Piatt.*)

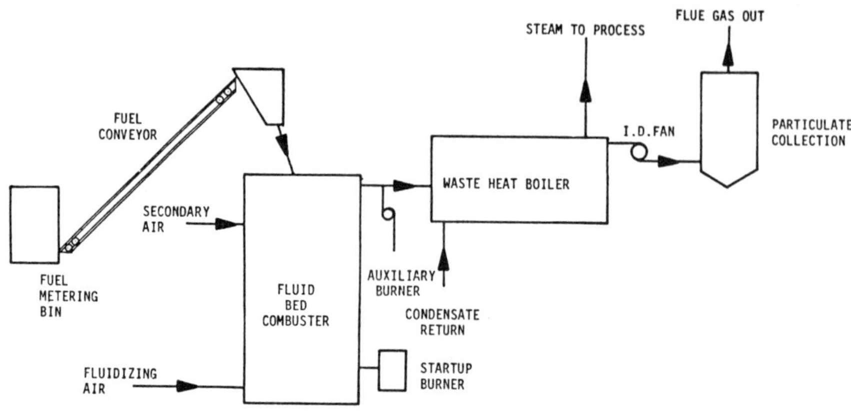

Figure 2-21. Fluidized bed wood combustion system.

chamber that has many holes drilled in the floor through which underfire air passes. The "bed" consists of small particles of sand, limestone, or other solid material. The bed is kept in suspension by forced air and is heated initially by an auxiliary fuel. When the bed reaches a temperature sufficiently high to ignite the fuel to be burned, the auxiliary fuel can be shut off and the solid fuel introduced. The turbulent mixing action of the hot bed material helps ensure that the fuel is burned completely. Fluidized beds have been looked at for coal burning because the bed material can be limestone. The limestone reacts with the sulfur in the coal to form a salt which can be removed from the bed, thus preventing the escape of sulfur compounds from the stack.

In wood burning systems, the fluidized-bed combustors have shown promise as devices capable of burning wet fuels or fuels of irregular sizes and shapes. In addition, other waste materials (carpet wastes, peanut shells, etc.) may be burned in conjunction with the wood. Several companies are actively marketing various fluidized bed systems and many units have been operated successfully in the forest products industry. Fluidized bed systems generally have a slightly higher (perhaps 10%) first cost and higher power requirements for the fans when compared to conventional systems. These disadvantages, however, are offset by the following:

- Short resident time (rapid combustion) in the fluidized bed is conducive to a faster reaction to changes in heat demand (load swings)—

there is rarely more than 30 seconds of fuel in the combustion zone at any time.

- Combustion efficiency is significantly higher in the fluidized bed, resulting in considerably less unburned hydrocarbons remaining in the ash.
- High ash content fuels are readily handled in the fluidized bed. Wide variations in moisture content, sizes, and heating values of fuel are easily accommodated. Several dissimilar fuels may be fed in series or simultaneously.
- Continuous operation is achieved with the fluidized bed. No need to take the unit off line to remove ash and debris. Optional bed cleaner, which operates while combustion continues, is available for dirty fuels.
- Reduced maintenance in the combustion chamber of the fluidized bed—no grates to be cleaned, repaired, or replaced. Control of combustion temperatures under 2,000°F is conducive to longer refractory life and eliminates slagging of ash.

York-Shipley. A wide range of standard models are available from 5 million Btu/hr to 120 million Btu/hr, as well as custom designs for unusual applications. Boiler conversions, complete steam generating systems in a variety of pressures and capacities, indirect hot-air systems with gas-to-air heat exchangers, and direct hot-gas systems for kilns and dryers are operating throughout the U.S. and Canada. A York-Shipley unit is shown in Figure 2-22.

Johnston Fluid Fire. The Johnston Boiler Company of Ferrysburg, Michigan, has obtained a patent license from Combustion Systems Limited in the United Kingdom to sell in the U.S. fluidized bed techniques developed in that country. The emphasis in Britain has been on coal firing using limestone as the bed medium, but the company has indicated successful operation on dry wood waste using sand as the bed medium.

The Johnston unit is available in sizes from 5,000 to 25,000 lb/hr of steam at pressures up to 300 psi. The Fluid Fire is a shop assembled package type unit transportable by rail (and by truck in the smaller sizes). As many components as possible are mounted on the unit prior to shipment to minimize field erection. Gas and oil auxiliary firing is normally provided and is necessary for startup. A multiclone collector

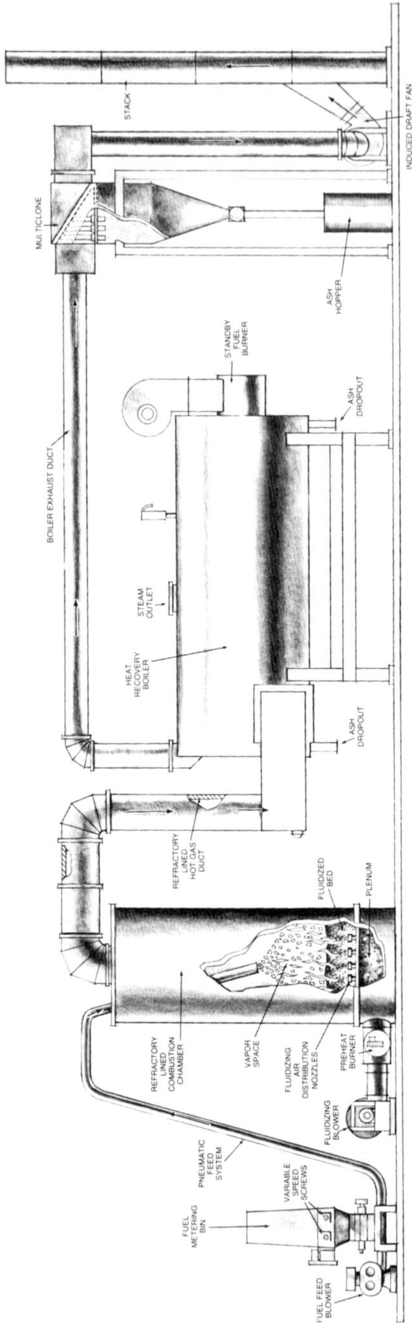

Figure 2-22. Fluidized bed wood combustor. (*courtesy York-Shipley Inc.*)

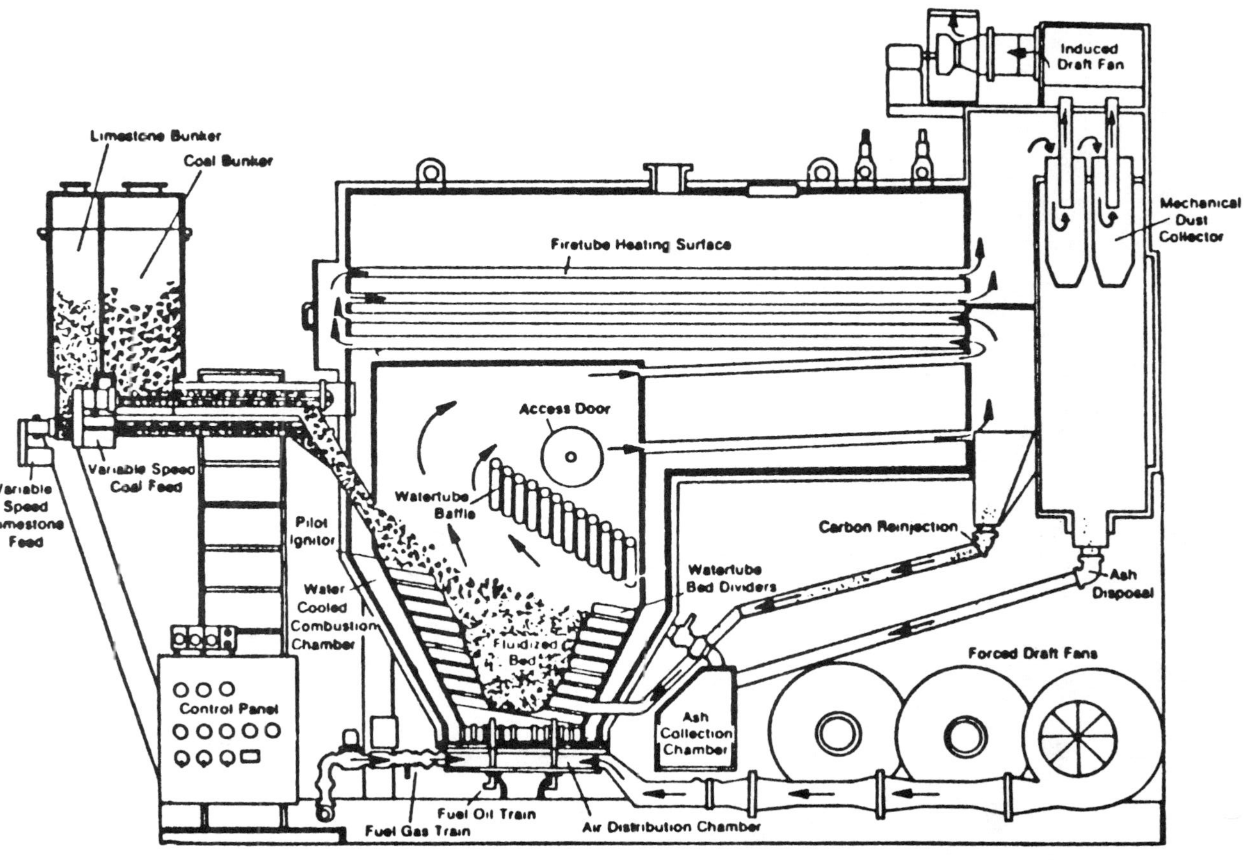

Figure 2-23. Package fluidized-bed boiler. (*courtesy Johnston Co.*)

is included in the package for air pollution control. A typical Johnston unit is shown in Figure 2-23.

Combustion Power Company. The Combustion Power Company is a division of the Weyerhaeuser Corporation and is located in Menlo Park, California. It is perhaps best known for its products for air pollution control, but it has been working for more than ten years on the development of fluidized-bed combustors. Some of its work has involved the combustion of waste fuels in a fluidized bed discharging into a gas turbine for power production, but the concept is not yet commercial. The company is interested in supplying fluidized bed units in capacities up to 100 million Btu/hr fueled by wood chips, industrial waste, or other low grade fuels. These combustors can discharge into boilers for steam generation or directly into dry kilns or veneer dryers for the lumber industry. In direct heat applications, Combustion Power achieves gas cleanup with one of its dry scrubbers or an air-to-air heat exchanger. As with other fluidized beds, some cleanup of particulate emissions is necessary.

At the Weyerhaeuser plant in Longview, Washington, Combustion Power constructed a fluidized bed to exhaust into an existing boiler. The fuel used was log-yard waste (55% moisture content) at a fuel feed rate of about 12,000 lb/hr. The combustion system is capable of producing 110×10^6 Btu/hr and is provided with fuel oil startup. A diagram showing the layout of this system is included in Figure 2-24.

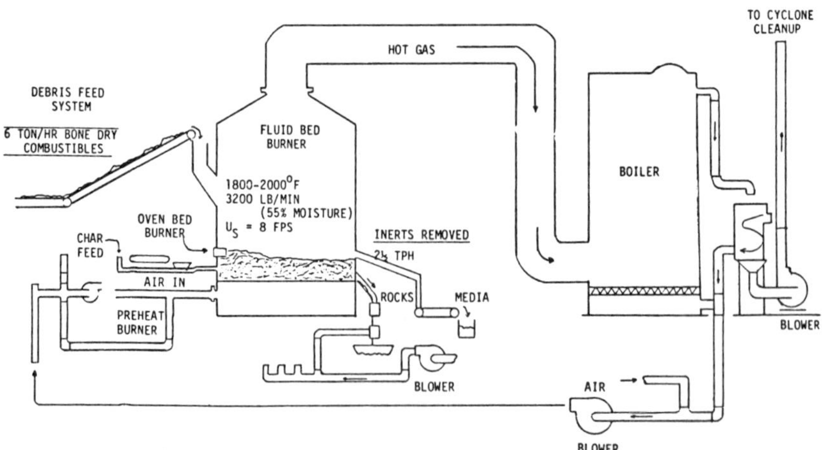

Figure 2-24. Fluidized-bed wood combustion system. (*courtesy Combustion Power Co., Inc.*)

Another industrial burner system was installed at Cosmopolis, Washington, to provide hot gas for a rotary dryer. The dryer is used to dry wood waste to fuel a conventional wood boiler. This unit was rated at 66 million Btu/hr and has been operated successfully.

It may be expected that the fluidized beds will require higher maintenance than conventional wood burners and that more highly skilled operators may be necessary to operate them. Research and development in fluidized-bed technology is very active at present. As all of the major large boiler manufacturers are investigating the use of fluidized beds to burn coal, much of the technology may filter through to wood systems. Other companies working on the perfection of fluidized-bed technology are included in Appendix 1.

PYROLYSIS SYSTEMS

Pyrolysis can be defined as burning at less than stoichiometric conditions. The process involves the physical and chemical decomposition of solid organic matter caused by the action of heat in the absence of oxygen. Wood can be pyrolyzed to produce charcoal, and coal can be pyrolyzed to produce coke. Due to the intense heat in a pyrolytic reactor, complex organic compounds can be broken down into simpler chemical products. These products include liquids, gases, and carbon char residue.

Various types of pyrolysis processes have been developed in recent years by many different companies. One of the goals sought by many of these companies is the production of useful fuels from municipal solid waste (MSW). The concept has merit, but dealing with large quantities of refuse is never easy, and no systems are considered completely commercial today. Limiting the feedstock to purely organic matter such as wood waste, agricultural residues, or the like, simplifies matters somewhat. Several of the systems that have shown promise for producing usable fuels from wood waste are discussed below.

At this time, pyrolysis units are to be considered (with some exception) "state-of-the-art" combustion devices; not proven reliable technology; and not generally recommended except where some of their special attributes dovetail to a plant's process (or fuel) requirements.

Tech-Air Corporation. Work on biomass pyrolysis was begun at the Georgia Tech Engineering Experiment Station in 1968, when a project was initiated to investigate methods for the disposal of agricultural residues without violating air pollution regulations. In particular, peanut hull disposal had become a problem for many Georgia growers, and it was decided to look for a method to use this potential energy rather than to just incinerate it. Several pilot units were built on the Georgia Tech campus, and pyrolysis experiments were carried out on various materials, including bark, sawdust, wood chips, cotton gin trash, various nutshells, automobile shredder wastes, and municipal wastes. The Tech Air Corporation was formed to help organize the development efforts. Two field test units (each with a nominal capacity of two tons per hour) were installed at a peanut shelling plant in Georgia. Char and pyrolytic oil were produced successfully with these units, and the experience gained led to the building of a commercial prototype plant in Cordele, Georgia. This prototype plant was operated intermittently for about two years, sometimes on a round-the-clock basis.

During this time, another pilot plant was built on the Georgia Tech campus designed specifically for the processing of municipal waste. This one-ton-per-hour unit was operated on light fraction, heavy fraction (with and without metals), whole garbage, sewage sludge, and shredded tires. Based on the results of this work, the American Can Company became interested in the process and bought the Tech-Air Corporation in 1975. At that time, two different efforts were initiated to carry forward the work of commercializing the wood-waste system and to establish a continuing research and development program at Georgia Tech in order to support the area of waste utilization.

During the eighteen-month operation at the Cordele plant, the char and oil produced was sold in the bulk char and fuel oil market. The system produced varying amounts of oil, gas, and char according to the control configuration. Energy content of the products, of course, depended on the feedstock, but, in general, the heating value of the char varied from 12,300 to 13,500 Btu per pound. Heating value of the gas produced by the pyrolysis system varied from 3,000 to 4,000 Btu/lb and the heating value of the oil was generally around 9,000 Btu/lb.

The char produced with the Tech-Air system has been used to make charcoal briquettes and has been experimented with in the laboratory to make activated carbon. The pyrolysis oil has been used as a fuel and has shown some potential as a chemical raw material. The oil has been sold commercially for use as a fuel in a cement kiln, a power boiler, and a lime kiln. The pyrolysis oil has also been successfully mixed with #6 fuel oil. The gas from the process has been used to fire an infeed dryer at the Cordele plant and, in addition, has been used to fuel an internal combustion engine. The gas had a heating value of up to 180 Btu per cubic foot.

At present, the future marketing efforts of the Tech-Air Corporation are not well defined, since most of the research and development work has been completed. Interested parties are encouraged to contact the individuals listed in Appendix 1. An overall view of the Tech-Air system is included in Figure 2-25.

Energy Resources Co. Inc. The Energy Resources Co., Inc. (ERCO) in Cambridge, Massachusetts, has been working for several years on the development of a pyrolysis system using various feedstocks including wood and agricultural wastes. Rather than using a plug flow, fixed bed such as Tech-Air and Enerco, ERCO accomplishes pyrolysis using a fluidized bed. The feed material is introduced into the fluid bed and partially burned to supply the heat for pyrolysis. The remaining solids are partially entrained and carried out of the fluidized bed. These solids contain char particles which are separated from the gas stream with a multiclone. The char thus collected can

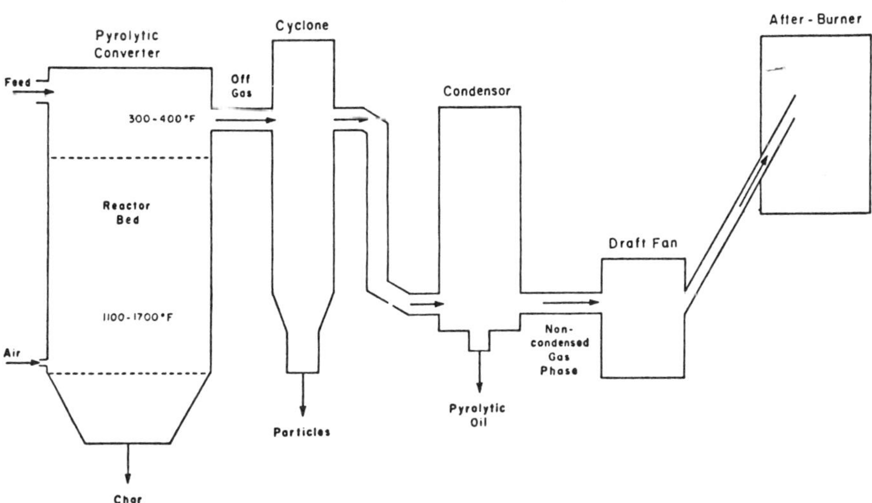

Figure 2-25. The Tech Air pyrolysis system.

be recycled to the fluid bed to produce a higher yield of gas and oil or it can be removed and sold as a feedstock for charcoal briquettes. The oil contained in the gas stream can be recovered by condensation or sent along with the gas to be used for process heating.

The fluid bed medium is sand and the manufacturer claims that fuel moisture contents of up to 50% can be handled. If this is the case, the ERCO system may have an advantage over the other systems already discussed in that it can burn more relatively wet fuels.

ERCO indicates that it is interested in supplying fluidized-bed pyrolysis systems that are capable of processing 200 dry tons per day of wood chips having a heat generating capacity of approximately 150 million Btu/hr. Larger units will use several reactors in parallel to achieve acceptable turndown ratios. It is interested in supplying turnkey installations with the help of an architect/engineering firm. A general view of the ERCO system is shown in Figure 2-26. A complete turnkey operation in Belle, Missouri, is in the startup stage at publication. In addition, there is a working commercial plant in Bakersfield, California, which fires its boilers on almond shells. This plant has a 20,000 lb/hr steam production rate.

Cost data for wood pyrolysis systems is difficult to obtain as the units under development are still largely in the prototype stages. As can be seen above, the systems tend to be complicated and expensive. The attractive feature of the pyrolysis systems is the possibility for sale of a high-value product (char), but the purchaser of a pyrolysis system should be quite certain about the existence of this market before he purchases one. Further information on suppliers of pyrolysis systems is included in Appendix 1.

WOOD GASIFIERS

Despite the many wood gasifier prototypes that have been placed on trial operation, installation of a wood gasifier (at this time) must be viewed partially as an R&D effort rather than a simple commercially proven steam plant modification. The indirect processes (gasification, pyrolysis, and liquefaction) represent a feasible technology which has been successfully applied in a widening circle of installations. Operational reliability of indirect combustion process equipment, however, has been uneven—mainly due to pilot plant/full size scale-up problems. Therefore, applications need to be evaluated on a case-by-case basis.

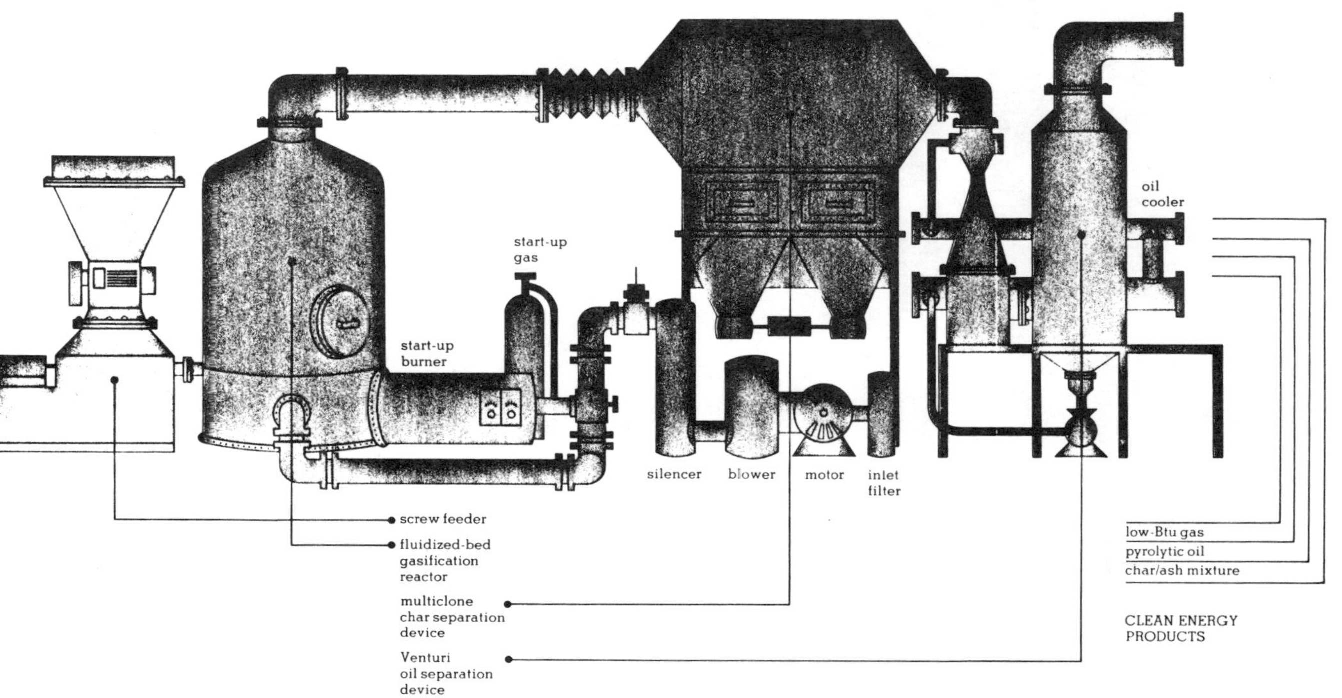

Figure 2-26. Fluidized bed pyrolysis system. (*courtesy Energy Resources Co., Inc. (ERCO)*)

The main problems associated with wood gasifiers involve undesirable constituents in the gas, including certain tars and acids. These have a tendency to condense in pipes and burners if the gas is allowed to cool or if the gas is not "scrubbed" as it exits the gasifier. These undesirable properties of the gas and the problems with grate slagging and burnout have been among the most formidable facing researchers.

The basic design for some classes of gas producers dates back to the late 1800's when wood gas was used to fuel engines for power generation and as a substitute for coal gas for lighting and industrial processes. Therefore, due to the age of the basic technology, it has been difficult for researchers and manufacturers to obtain patents as "new" developments. Thus, many small concerns are reluctant to divulge the specifics of systems they are developing. Wood gasification systems available today cannot be considered fully "commercial" in the sense that an industrial plant could buy a unit "off the shelf" and install it in a plant and expect it to operate nearly automatically for 7,000 to 8,000 hours per year. The state-of-the-art is not yet to that point and, in fact, some units have been placed in customers' hands before the serious operational problems have been resolved.

The several concepts being developed for gasifiers at present include updraft gasifiers, downdraft gasifiers, cross-flow gasifiers, and fluidized-bed gasifiers. Before considering specific systems in further detail, Figure 2-27 will be used to show some basic principles of gasification. An updraft unit is shown, but the same basic principles are involved in all units.

The gasifier vessel is filled with wood chips and a small fire is started on a grate above the ash hopper. Just enough air is introduced into the unit to support a glowing bed of coals. Steam may be injected into the bed to promote the formation of hydrogen (H_2) and to control grate temperature. In the reduction zone, the hot gases from the combustion process are partially reduced to carbon monoxide (CO), which is a medium heating value gas. The products then pass into a pyrolysis zone where the infeed first gives up its volatile gases. At this point, many of the materials which cause problems in the gasifier such as tars and acids are produced. In the top section of the updraft unit, a drying process occurs wherein the exiting product gases give up some of their heat to the infeed. The gas is drawn off at the top of the unit and piped to a burner or engine for final combustion. Other methods of gasification differ in regard to where the reactions occur, but the gasification pro-

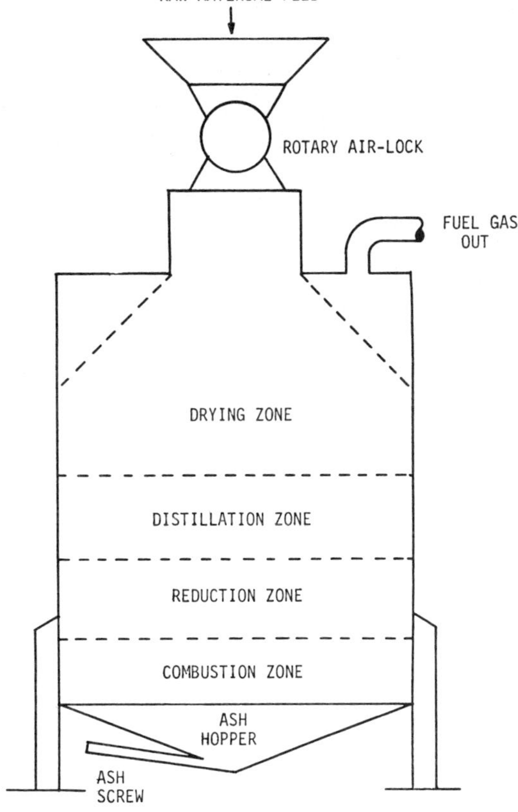

Figure 2-27. Updraft wood gasifier.

cesses are basically all the same. Due to the various gasifier configurations being developed by different manufacturers, the gases produced differ somewhat in their contents. Typical heating value of gas produced in wood gasifiers is 150 Btu/cu ft.

The Davis Gasifier. The "University of California at Davis" gasifier is a downdraft configuration as shown in Figure 2-28. This unit was designed for the gasification of agricultural wastes, but has performed on wood residue as well. Some development problems have not been completely solved, and sources indicate that ash removal and slagging have presented problems for long term operation. Several variations have been built and a spinoff company called the Biomass Corpora-

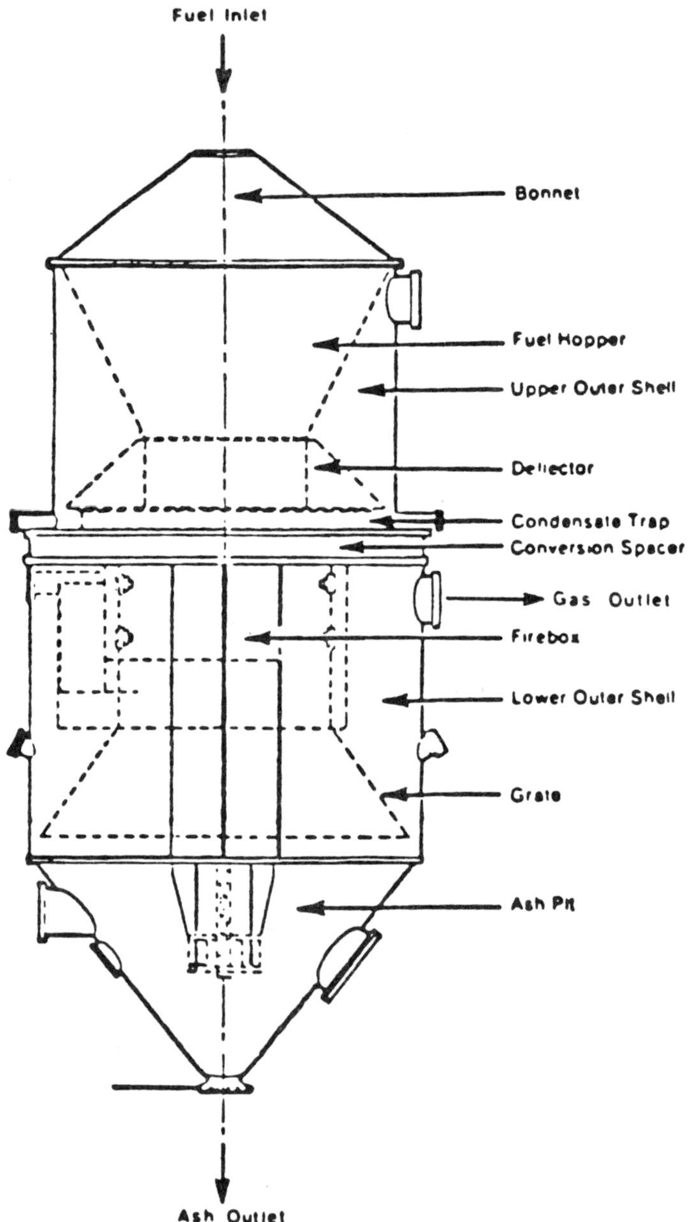

Figure 2-28. University of California-Davis gasifier. (reprinted from: Goss, J.R. and Williams, R.O., *Walnut Shells: Replacement for Natural Gas?*, published in Chilton's *Food Engineering Magazine*, Sept 1977.)

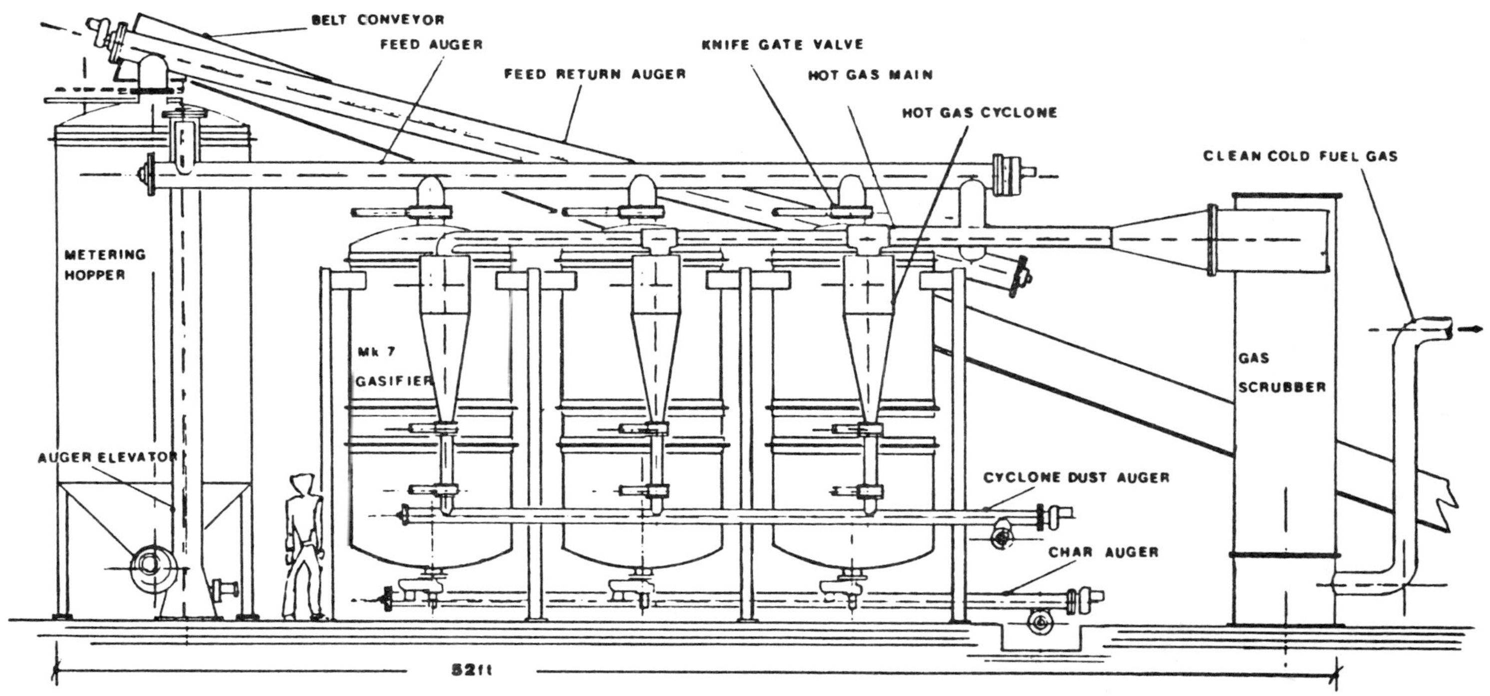

Figure 2-29. Three-unit gasifier set. (*courtesy The Biomass Corporation.*)

tion has been formed to develop commercial applications. Figure 2-29 shows a unit produced by the Biomass Corporation. Most fuels used in these units have been quite dry. The largest units in operation thus far are in the 10 MMBtu per hour range.

The Forest Fuels Gasifier. The Forest Fuels Company of Keene, New Hampshire, has been developing a cross-flow type inclined grate wood gasifier for several years now, and several units have been placed in the field. A cross-section view of the latest configuration is shown in Figure 2-30. As can be seen from the figure, the primary air supply passes through the fuel bed which is gravity-fed to an inclined traveling grate. Secondary air is introduced at the burner where the combustion process is completed. The largest unit known to be operated in the field at this time is about 15 MMBtu/hr, although the manufacturer is interested in supplying units in the range of 1 to 18 MMBtu/hr.

Duvant Gasification System. Moteurs Duvant, a division of the Renault Corporation located in Valenciennes, France, is marketing an electric power generation system using biomass as the feedstock for a downdraft gasifier. Units presently in operation have been operated on coconut husks, wood waste, corn cobs, and other organic wastes. The manufacturer lists five units in commercial operation. The gases produced from the biomass pyrolysis are used to power a conven-

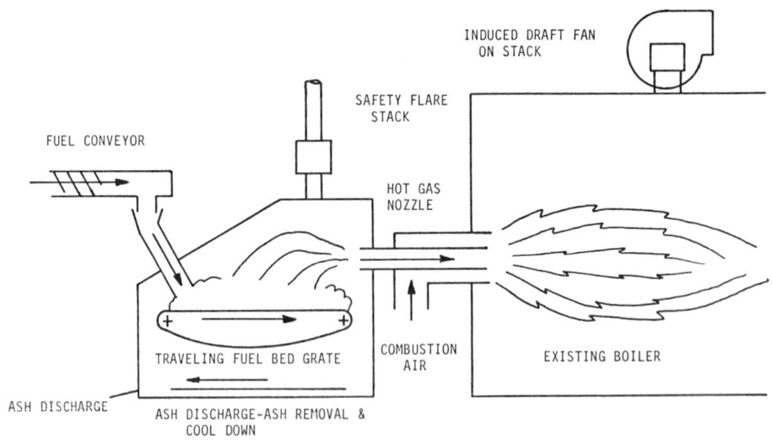

Figure 2-30. Cross-flow wood gasifier. (*courtesy Forest Fuels Inc.*)

tional diesel engine which drives an electric generator. Duvant is interested in supplying power outputs in the range of 120 to 740 kw. The engine does not run entirely on low Btu gas, but is co-fired with as little as 10% fossil fuel. The engines are derated by about 15% when the switch is made from 100% fossil fuel to 10% fossil fuel.

An important part of the gasification system is the gas scrubber which is used to remove particulates from the gas stream in order to prevent damage to the engine. Presumably, unscrubbed gas could be burned directly in a boiler or other burner, but the manufacturer does not indicate if this has been done.

These systems would appear to be especially suited to remote areas which do not require large electrical supplies and have large biomass reserves. In fact, most of the installations have been made in underdeveloped countries.

Duvant is interested in turnkey installations, and an outline drawing of a typical system is shown in Figure 2-31.

There are many other wood gasification systems under development, but most are similar to the units discussed above. A more complete listing of the companies engaged in research in this field is included in Appendix 1.

In general, the main barriers to full scale commercialization of gasifiers include the following: (1) lack of long term operational experience by current manufacturers; (2) potential problems in piping and burners resulting from tars and other liquids in gases; (3) slagging of grates due to ash contained in wood; and (4) requirements for operator attention. Indications are good that these problems will be solved by several manufacturers over the next few years.

One of the most attractive aspects of wood gasifiers is price. As mentioned earlier, packaged wood boilers may cost four to five times as much as the natural gas/fuel oil boiler that has been the mainstay of the commercial market for the past thirty years. Cost figures obtained by the Solar Energy Research Institute and independently by Georgia Tech indicate that gasification systems can be retrofitted to an existing gas/oil burner for substantially less than the cost of a new wood system. Indeed, the major portion of the cost of a gasification system will be for the wood handling, conveying, and metering components. Thus, the industrial plant owner who has a gas/oil boiler with significant useful life remaining will have an attractive alternative to an entirely new system. The cost curves in Chapter 10 illustrate this point more clearly.

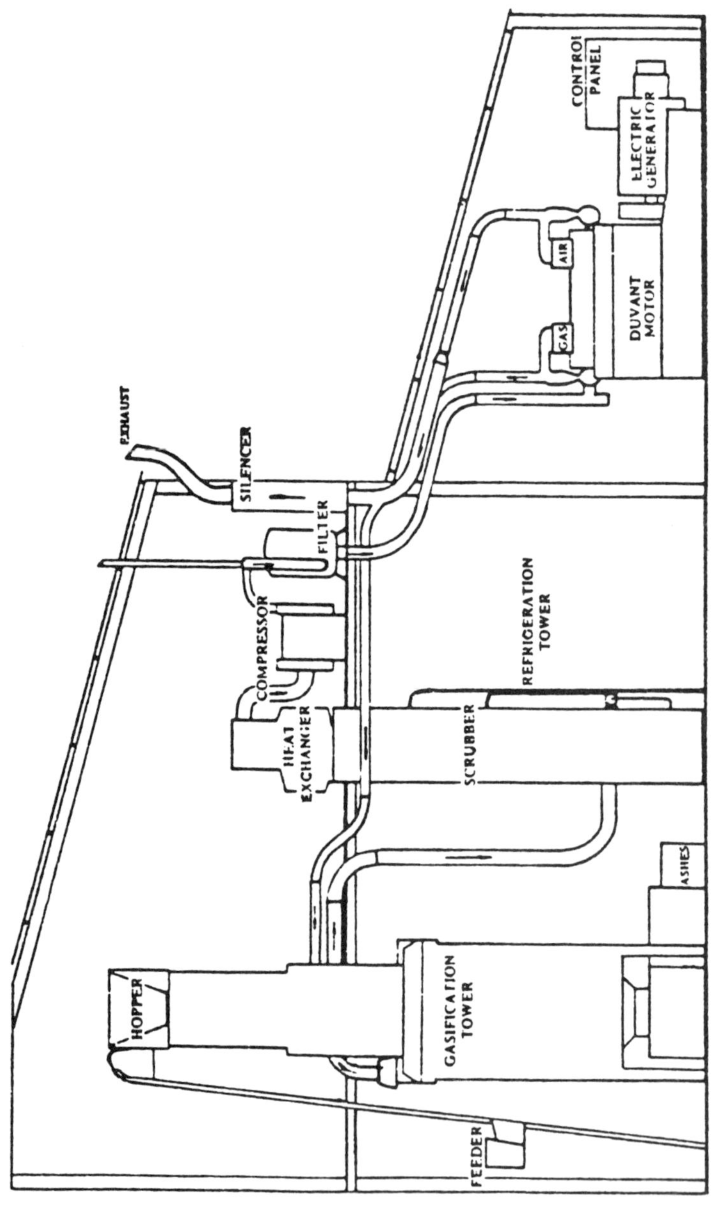

Figure 2-31. Wood gasification system for electric power generation. (*courtesy Moteurs Duvant.*)

OTHER COMBUSTION SYSTEMS

In addition to the wood burning systems already discussed, there are systems available which do not fit conveniently in the above categories. This hardware is discussed below.

Lamb-Cargate. The Lamb-Cargate Company of New Westminster, British Columbia, specializes in the supply of hardware for the forest products industry. For several years, they have been working on a system for burning wet wood residue and chips which employs features from several of the categories already discussed. This system is called the Lamb-Cargate ''Well Cell'' and is illustrated in Figure 2-32. The unit functions partially as a gasifier and partially as a cyclone-type burner. The Wet Cell contains two chambers. Wood residue is fed with an underfeed stoker and is forced up into the center of the horizontal pinhole grate where it forms a conical pile in the primary chamber. Underfire air at 500°F is blown through the bed and the primary chamber functions as a partial gasifier. The volatiles and fuel gases are driven off and then partially combusted in the top of the primary chamber. Combustion is completed in the secondary chamber, and gases exit the top of the unit at high temperature. The manufacturer claims that the unit is capable of burning 65% moisture content (wet basis) wood, and the unit has also been operated on drier wood down to 30%. Depending on the moisture content, the output temperature is reported to be 1700 to 2300°F. The output of the unit can thus be ducted to a waste-heat boiler or various types of dryers. An early production Wet Cell has been fitted to a lumber dry kiln in northern British Columbia and is reportedly operating satisfactorily. Lamb-Cargate is presently testing a 60 million Btu/hr Wet Cell to be retrofitted to a rotary lime kiln for a paper mill in Vancouver. Auxiliary heat input is necessary to reach the temperature range (2300 to 2700°F) used in this application. If successful, these units could conceivably be fitted to many other types of rotary dryers, including mineral dryers and perhaps even cement kilns.

Control of the Wet Cell is accomplished by controlling the amount of air admitted to the lower chamber. This amount is automatically regulated so that the mixture of volatiles and water vapor passing to the second chamber is at an optimum temperature, generally below 1600°F. The secondary control then admits the necessary combustion

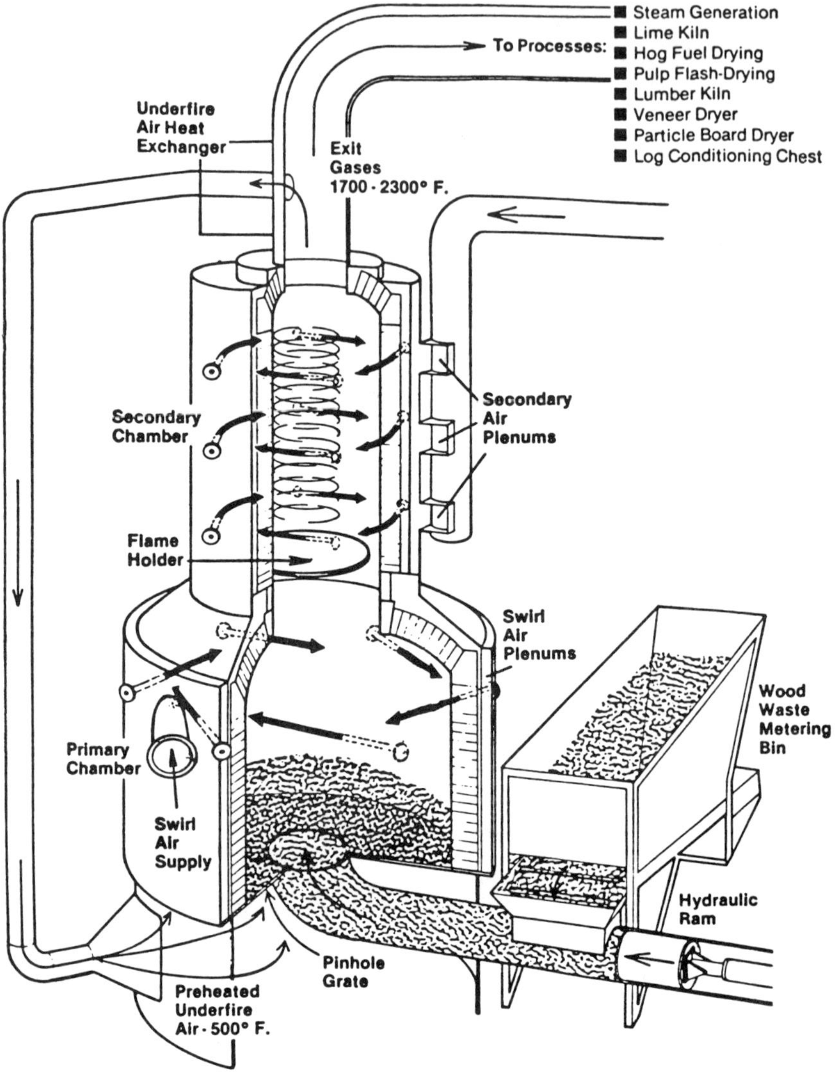

Figure 2-32. Lamb-Cargate wet cell. (US Patent 885-377 Can. Patent 322-921.)

air so that the products of combustion leave the upper chamber at a steady temperature.

Prototype units were in the 25 MMBtu/hr range, but Lamb-Cargate has units under development which will generate up to 75 MMBtu/hr. Particulate emissions are reported to be ''low,'' but some mechanical collection may be required. Cost estimates are not complete due to the early stage of development. It would appear that this system would be cost competitive with a wood gasifier for a given output.

3

Wood Fuel Storage and Handling

WOOD FUEL PROPERTIES

For material handling purposes, whole green tree chips, bark, and mill residues act (to a large degree) more like earth than a free flowing substance. For wood fuel, the varying characteristics of concern during storage and handling are its moisture content and size.

Wood fuel can be a waste product of lumber or furniture manufacturing. It can also come from waste generated by conventional harvesting techniques or it can be directly harvested from the forest. Most available wood fuels are green. Dry wood fuel is waste from manufacturing (such as furniture operations) that uses kiln dried lumber. The choice between dry or green wood fuel for a particular location is decided on the basis of availability and cost. One consideration is that combustion equipment for green fuel can also burn dry fuel, but the reverse is seldom true. It is believed that most systems will be based on green fuel and may burn dry fuel as it becomes available.

The physical size of wood fuel is also important and affects handling, storage, on-site fuel preparation, combustion systems, and emissions. There are presently no official standards for wood fuel. Current practice is to analyze local availability of wood fuel and design the plant system accordingly. General classifications are sawdust up to ⅛ in., whole green tree chips up to 2 in., and wood waste—a mixture of sizes from sawdust to pieces several feet long. Wood waste may require size reduction. This can be accomplished by purchasing prehogged material or, in the instance of large systems, by hogging on-site as a part of the plant's fuel handling system.

Another general classification of wood fuel that is available in certain areas is ''bark and shavings,'' derived from debarking of logs. Depending upon the maximum size of the bark and shavings, this fuel may require size reduction. Bark fuels tend to be high in silica (sand) content. Sand enters the bark in two ways: (1) in coastal regions sand is imbedded into the surface layers through wind transport; (2) during skidding (especially in rainy seasons) sand is picked up by the surface layers. The high silica content (non-combustibles may reach 14%) creates problems in clogging of boiler grates, erosion of fuel and ash handling systems, heat exchanger wear, and increased particulate emission.

Wood fuel for industry can be purchased on several bases, but the common method is by ton weight. As the wood fuel industry progresses, it is expected that standards will evolve that reflect actual heat content. However, this lack of standards is not unusually critical in that the various green wood materials range plus or minus 5% of 4,000 Btu per pound. For dry wood, more care must be exercised because the percentage of drying affects the heat content. For example, kiln-dried material may have a 10% moisture level and contain 7,200 Btu per pound.

Manufactured wood fuel for industrial use is increasing in availability. A promising technique that currently has limited commercial distribution is densification. Wood is dried, reduced to small particle size, and then densified by extrusion. This process produces a product in the form of pellets or cubes. The bulk density is about 35 lb/ft^3 compared to about 23 lb/ft^3 for wood chips. The uniform product will allow storage, handling, and combustion equipment to be designed to its specific characteristics. Combustion thermal efficiency can be 15% higher than is obtainable with green wood, and the capital cost of the combustion equipment can be reduced. The costs of densification are significant, however, so the place of manufactured wood fuel is not clear at this time. Typical selling prices are in the range of $50/ton, three times the cost (on a Btu basis) of wood residue.

In order to plan the size of wood yard required, the boiler fuel requirements must be determined.

TABLE 3-1. Wood Fuel Properties.[40]

WOOD FUEL	MOISTURE CONTENT, WET BASIS	HEATING VALUE (BTU/LB)	BULK DENSITY (LBS/FT^3)
Whole tree chips	50%	4,000	24.0
Green sawdust	50%	4,000	20.0
Dry planer shavings	13%	6,960	6.0
Dry sawdust	13%	6,960	11.5
Wood pellets	10%	7,200	35.0

Using Table 3-1, the calculations for determining the quantity of fuel required are:

$$\text{Quantity of fuel required} = (\text{steam demand}) \times (\text{enthalpy of steam}) \times \\ 1/\text{HHV of wood} \times 1/\text{boiler efficiency}$$

where

Steam demand is in lb-steam/hr
Enthalpy of steam (at 150 psi; 220°F feedwater) = 1000 Btu/lb-steam
HHV of the specific wood fuel: as specified
Boiler efficiency = 0.65 (typical for green wood fuels)

For a 1,000 hp boiler burning whole tree chips (50% M.C. wet) at 65% efficiency, the hourly fuel requirement is:

$$(1,000 \text{ boiler hp/hour}) \times (34.5 \text{ lb-steam/boiler hp}) \\ \times (1,000 \text{ Btu/lb-steam}) \times (1 \text{ lb of wood})/(4,000 \text{ Btu/lb}) \\ \times (1 \text{ ton}/2,000 \text{ lb}) \times (1/0.65 \text{ boiler efficiency}) \\ = 6.6 \text{ tons/hour (of green whole tree chips)}$$

For some commonly used industrial boilers, Table 3-2 is calculated using green wood chips as the fuel and a boiler combustion efficiency of 65%. This table assists in calculations necessary for sizing the wood fuel handling and storage facilities.

TABLE 3-2. Wood Fuel Quantity Requirements.[17]

| | STEAM OUTPUT | WOOD CONSUMPTION | | | TRUCKLOADS/DAY |
BOILER HP	(LB/HR)	TONS/HR	TONS/24 HRS	FT³/HR	(23 TONS/LOAD)
100	3,450	0.67	16.08	55.83	≈ ¾
200	6,900	1.35	32.16	111.67	≈ 1½
300	10,350	2.01	48.24	167.49	≈ 2
400	13,800	2.70	64.32	223.33	≈ 3
500	17,250	3.35	80.40	279.17	≈ 3½
600	22,500	4.02	96.48	334.98	≈ 4
700	24,150	4.69	112.56	390.81	≈ 5
800	27,600	5.40	128.64	446.67	≈ 6
900	31,050	6.03	144.72	502.47	≈ 7
1,000	34,500	6.60	160.80	558.33	≈ 7

Notes: 1. Fuel is green wood chips.
2. Boiler combustion efficiency assumed to be 65%.

RECEIVING

There is no one optimum method by which wood fuel can be received and unloaded at an industrial plant. It is necessary to evaluate specific conditions at each site. The capabilities of prospective wood fuel suppliers also influence the decision. Economics is a major consideration, and the lowest cost method that meets the particular needs should be selected.

Figure 3-1[22] illustrates four receiving methods that are widely used, and Table 3-3[22] shows estimated costs per year.

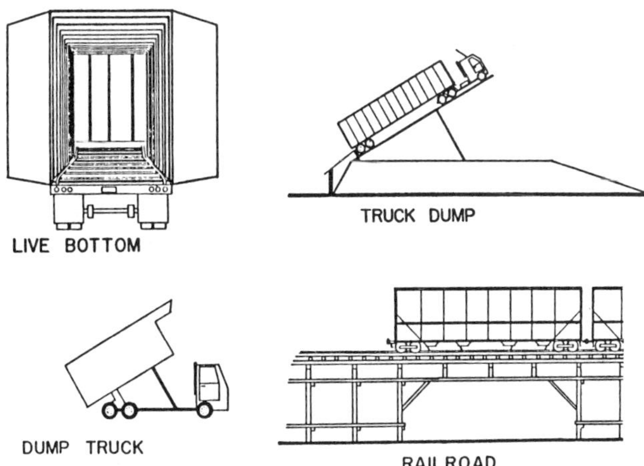

Figure 3-1. Wood fuel receiving methods.

TABLE 3-3. Comparisons Between Unloading Alternatives.[22]
(estimated cost per year)

BOILER SIZE			LIVE BOTTOM	TRUCK	FRONT END	SCOOP
HP	LB/HR	TONS/YR	VAN [1]	DUMP [2]	LOADER [3]	ROVEYOR [4]
(100 hp)	3,450	4,900	$ 1,000	$13,000	$ 4,960	$ 9,350
(1000 hp)	34,500	49,000	$10,000	$14,000	$20,260	$12,950
(3000 hp)	100,000	142,000	$29,000	$16,000	$51,260	$20,450
(7500 hp)	250,000	355,000	$73,000	$33,567*	—	$46,400*

Note: Interest Rates are assumed to be 10%.
*Two unloading devices required.
[1] Haul is 30 miles one way, 22 tons per load, live bottom vans add $.15/mile.
[2] Installed cost of truck dump is $110,000, life is 15 years, unloading time is 15 min.
[3] Cost of loader is $20,000, unloading time is 45 minutes, labor is $5/hour, operating costs are $5/hr, tractor life is 10 years.
[4] Cost of Scoop-Roveyor is $55,000, life is 10 years, unloading time is 15 min, labor is $5/hour, and operation is $2/hr.

- *Live Bottom Van*. The live bottom van is a self-unloading trailer with a conveyor incorporated into the floor. A typical van is 40 ft long, 8 ft wide, and 13 ft 6 in. high and will transport about 23 tons of material per load. The conveyor travels at the rate of six feet per minute, permitting the van to be unloaded in under 10 minutes. Live bottom vans cost approximately $28,000—which is twice the cost of standard open top trailers of the same size. The costs incurred to operate the van are reflected by increased transportation charges on the order of 10% to 15%. The advantage to fuel purchasers of live bottom van deliveries is that no costly unloading device is required on-site. During discussions with companies presently in the waste wood business, several expressed the opinion that the live bottom van will be the method of choice for many industrial plants that require six deliveries per day or less.

- *Truck Dumps*. Truck dumps are devices installed at the plant site that elevate and unload trailers. Most are hydraulically operated. They come in two sizes—one handles both tractor and trailer and the other handles the trailer only. A truck dump with a 40-foot platform to handle a standard trailer costs about $35,000; a dump with a 60-foot platform to handle both the tractor and trailer costs around $84,000. A complete (installed) installation including a live bottom receiving hopper can exceed $120,000. The major advantage of a truck dump is that it can quickly unload widely available standard 40-foot vans. Turn-around time can be 20 minutes or less. Truck dumps are most practical for larger installations that require more than 10 deliveries per day.

- *Railroad Delivery*. Receiving wood fuel by railroad is practical under the right circumstances. Railroad shipment can be the lowest cost transportation when shipping distances exceed 70 miles. If a railroad siding is available and the local area wood fuel is in short supply, then railroad shipment should be evaluated.

- *Dump Trucks*. Dump trucks are sometimes used for wood fuel delivery. Though usually of smaller capacity than standard 40-foot vans, they can be utilized in specific circumstances for short-haul situations.

- *Other*. One semi-automatic method of truck unloading is the "Scoop-Roveyor" device by Morbark of Winn, Michigan. A "Scoop-Roveyor" resembles a coal seam miner and actually digs its way into the rear of a chip van, emptying the wood in 10 to 15 minutes. An operator to control it is stationed at the collection end. The cost of this device is from \$50,000 to \$55,000 depending on options and capacity. Scoop-Roveyors may be the most economical unloader for plants of medium-size.

  Small installations not having specialized unloading devices could employ a front end loader for this purpose. Front end loaders can unload trailers in around 30 to 45 minutes. When the tractor operation and labor costs are considered, this method can be justified economically only in a few small-size operations. Unless the operator is skilled, damage to the trailer can occur.

STORAGE OF WOOD FUEL

Storage of wood fuel is one of the first considerations in the design of a wood fuel facility. Often a significant land area is required, and the location of available storage area relative to the combustion burner location affects the fuel handling system design and costs.

The size and type of wood storage required varies with the requirements of a particular plant. Among the factors to be considered are:

- Heat energy requirements of the plant
- Land area availability and its location
- Fuel moisture content
- Fuel preparation requirements
- Reliability of wood fuel supply system
- Weather severity

All factors should be evaluated and considered in the storage system. The actual storage is usually accomplished by open storage, covered storage, or by silos. Figure 3-2[22] provides schematic views of typical storage methods.

Open Storage

Open storage is an area that is not covered for protection from rain or other precipitation. It can be the only storage needed, but is most often

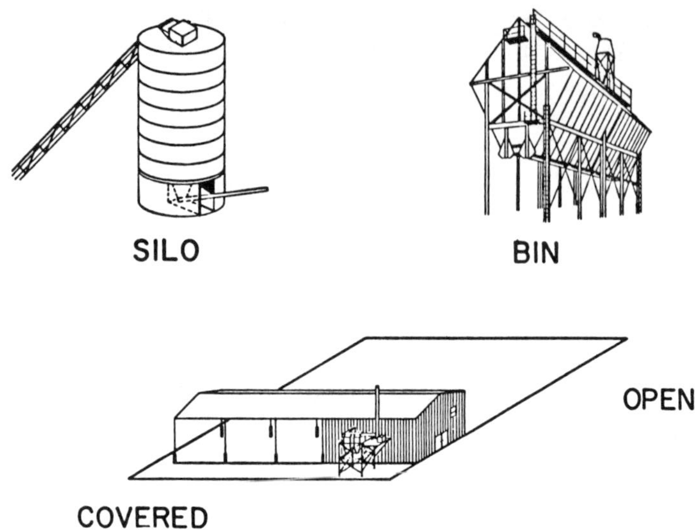

Figue 3-2. Wood fuel storage methods.

used for a 10- to 30-day supply to handle interruptions in the normal fuel delivery system such as inclement weather or strikes. The quantity of material storable on a given site is limited to the height obtainable by the stacking and retrieval equipment.

It is recommended that the open storage area be paved, preferably with concrete, to prevent the accidental scraping up of dirt with the fuel which can cause problems with the handling and combustion equipment. The area should be pitched to allow for drainage and should not be located in a flood prone area.

All wood fuels undergo losses in net available energy as a function of storage time. (See Figure 3-3[23].) "Lifo" (last-in, first-out) rather than "fifo" (first-in, first-out) fuel utilization procedures are recommended.

The primary cause of depletion of *available energy (Btu/lb as fired)* of openly stored wood fuel is an increase in the surface layer moisture content due to precipitation. Once saturation of the surface volume is achieved, future weather conditions will not affect moisture content in this outer zone. Due to the high packing density and minimal air flows through most piles, the surface volume—within about 1½ to 2 ft of the surface—will remain saturated. Piles constructed in shapes having large

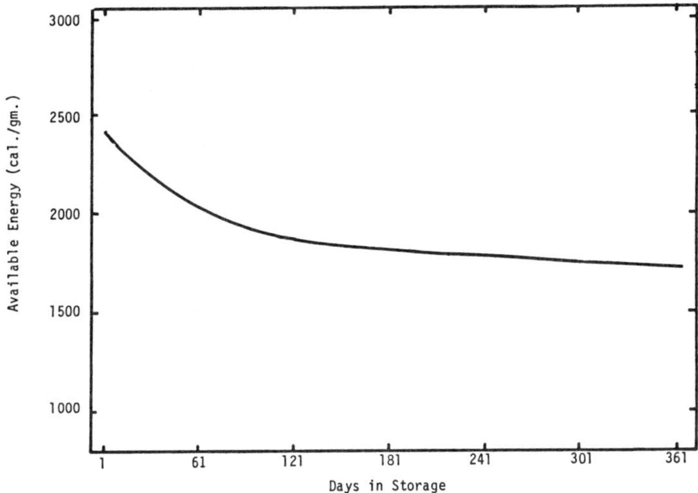

Figure 3-3. Available enery as a function of storage time for wood chips. (*courtesy of M. L. Curtis; from: "The Effects of Outside Storage on the Fuel Potential of Green Hardwood Residues" 1980, M.S. Thesis, Virginia Polytechnic Institute and State University, Blacksburg Virginia 24061.*)

surface volume/total volume ratios (i.e., rectangular piles) usually display net increases in average moisture content through the pile. Construction of conical piles will minimize surface volume/total volume ratios and consequent increases in net moisture content and will thus minimize any decreases in available energy.

The secondary cause of available energy depletion is loss of volatiles through evaporation—accounting for perhaps 15% of loss in available energy. Such losses are tempered by the poor air flow characteristics typical of most piles (Figure 3-4[23]). This "chimney effect" contributes to high internal pile temperatures (Figure 3-5[70]). Of course, the high internal pile temperatures can contribute to significant moisture content decreases in the central zones (Figure 3-6[70]).

Strongly influencing the natural drying potential of stored wood fuel is the packing density of the pile. The packing density is the configuration of the discrete wood chips and/or grains in the as-constructed state. Piles constructed with bulldozers or like equipment will have higher packing density and inferior air flow characteristics relative to those constructed by the gravity method. It can be expected that gravity con-

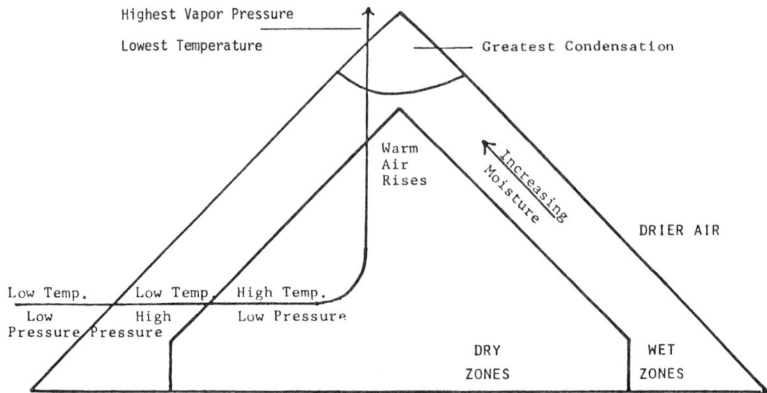

Figure 3-4. Chimney effect in storage piles. (*courtesy of M.L. Curtis; from: "The Effects of Outside Storage on the Fuel Potential of Green Residues," 1980, M.S. Thesis, Virginia Polytechnic Institute and State University, Blacksburg, Va.*)

structed piles will maximize the drying effects of naturally occurring high internal pile temperatures.

Investigations are currently underway at Georgia Tech to determine if there may be economic justification for covered long term storage. The results of these investigations are expected subsequent to publication of this handbook.

Another aspect of long term storage important to the plant engineer is the well-documented decrease in the pH of stored wood fuels (Figure 3-7[70]). The pH of stored wood is commonly in the acidic range of 4 to 5. This can be important in the design of the following portions of wood systems:

1. *Concrete slabs or structures* in intimate contact with the wood fuel (use of acid-resistant concrete mix and epoxy coated reinforcing bars is effective here).
2. *Any steel portions* of the material handling system—i.e., augers, ductwork, conveyor superstructures, etc.—can be expected to corrode at an accelerated rate if anti-corrosion/preventive maintenance procedures are not instituted.

Long term open storage of green whole tree chips, residues, and bark necessarily entails comprehensive safety procedures. The highest internal pile temperatures are found in bark piles, followed respectively by green whole tree chips and mill residues/sawdust (Figure 3-5).

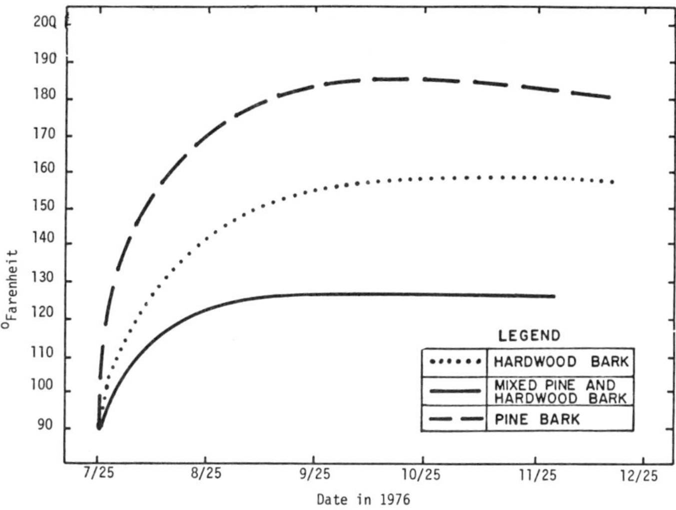

Figure 3-5. The maximum internal temperatures measured within experimental storage piles of residues as a function of storage time. (*courtesy of M.S. White from: "Bulk Storage Effects on the Fuel Potential of Sawmill Residues," 1978, Forest Products Journal, 28(11): 24–29.*)

With due consideration of safety, space, and cost efficiencies, the following storage procedures are recommended.

- *For Sawdust and Mill Residues*
  1. When natural drying is not a criteria, it is safe and practical to construct large (in areal extent), high piles compacted with heavy equipment such as loaders, bulldozers, etc. Installation of one thermocouple (type "T"—copper constantin) per six feet of pile height—in cold climates (see discussion below)—will minimize fire hazards. Should pile temperature exceed 170°F, danger of fire is imminent and the pile should be dismantled.
  2. When natural drying of stored fuel is the desired goal, build gravity-constructed piles. Thermocouple installation is recommended; however, elevated temperatures should not be expected.

- *For Whole Green Tree Chips and/or Bark*
  As the percentage of bark in a pile increases, so will internal temperatures. The occurrence of high pile temperatures in bark—and to a lesser extent whole tree chip—piles is well documented (Figure 3-

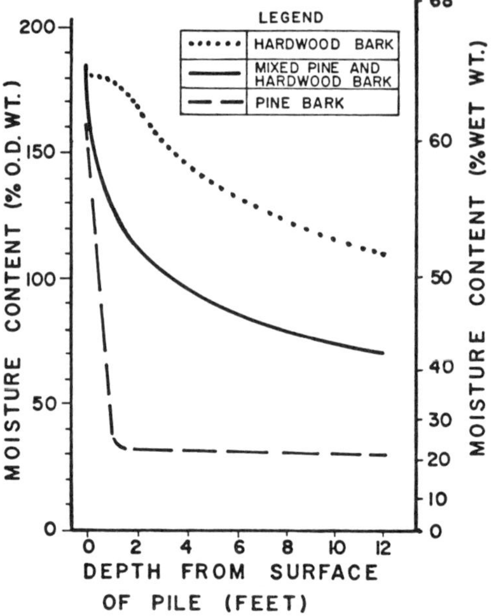

Figure 3-6. Moisture content as a function of depth of green bark and sawdust piles. (*courtesy of M.S. White, from: "Bulk Storage Effects on the Fuel Potential of Sawmill Residues," 1978, Forest Products Journal, 28(11): 24–29.*)

5 and research in progress by Georgia Tech). Therefore, long term storage piles of bark and/or whole green tree chips should be gravity constructed, *not* compacted with heavy machinery, and instrumented with thermocouples—one for every six feet of pile height. Again, should pile temperatures exceed 170°F, danger of fire is imminent and the pile should be dismantled.

The greatest danger of fire exists under conditions that restrict and/or eliminate air flow through the pile (and consequent internal pile heat dissipation). A well-documented cause of spontaneous fires occurs during the winter months when a layer of ice can form over the outer pile surface. During such periods, internal pile heat cannot be dissipated and fires often result. High concentrations of bark fines at the pile surface can also replicate this condition. Close observation of internal pile temperatures during these periods will help ascertain imminent danger.

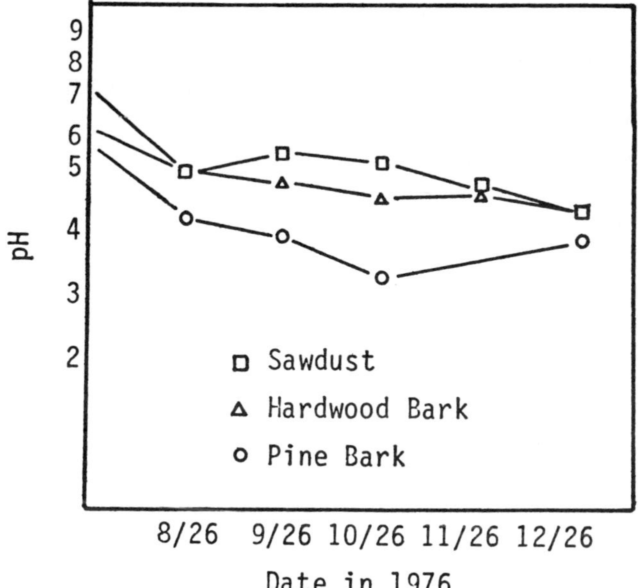

Figure 3-7. The average pH of residues within the experimental storage piles as a function of storage time. (*courtesy of M.S. White from : "Bulk Storage Effects on the Fuel Potential of Sawmill Residues," 1978, Forest Products Journal, 28*(11): 24–29.)

For an exhaustive state-of-the-art discussion on the above subject, the reader is referred to "The Effects of Outside Storage on the Fuel Potential of Green Hardwood Residues," by Michael L. Curtis.[23]

As stated, the storage volume per unit area is determined by pile height, and the total weight is obtainable by multiplying the volume by the specific weight of the particular wood fuel. As an example, if whole tree chips at 23 lbs/ft^3 are stacked to an average height of 12 ft, then 138 tons can be stored per thousand square feet of area. The cost of paved open storage depends upon local conditions (amount of area preparation necessary), but an average cost of $1.70 per sq ft can be used for preliminary estimating purposes.

Covered Storage

Wood storage systems frequently contain a covered storage area, using an open-side metal building located adjacent to the combustion equipment. It is sized for a three- or four-day storage capacity. Many of the same considerations considered under the discussion of open storage apply to covered storage.

Fuel to be immediately burned is taken from this area. The cover prevents soaking by rain. Under normal operation, most of the fuel can be delivered to the covered storage area and consumed without going through the open storage area.

Covered storage is necessary for dry fuel, such as fuel from kiln dried lumber or prepared, densified wood fuel. For estimating, Figure 3-8A[22] shows storage capacities for whole tree chips and the typical costs of covered storage sheds.

Silos

Silos are used to store wood fuel under certain conditions. For the non-forest industry, it is expected that silos will be considered when

1. Fuel requirements are relatively small, about one fuel delivery per day or less.
2. Automated fuel feed is highly desirable.
3. Site location prevents the use of open storage.

Silos come in a variety of diameters, heights, and volumes; and they can be manufactured of metal, poured concrete, or staved concrete.

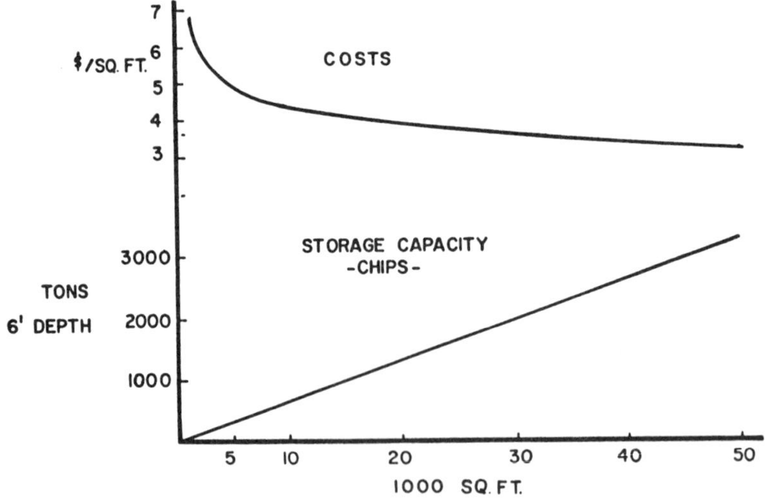

Figure 3-8A. Covered storage costs and capacities.

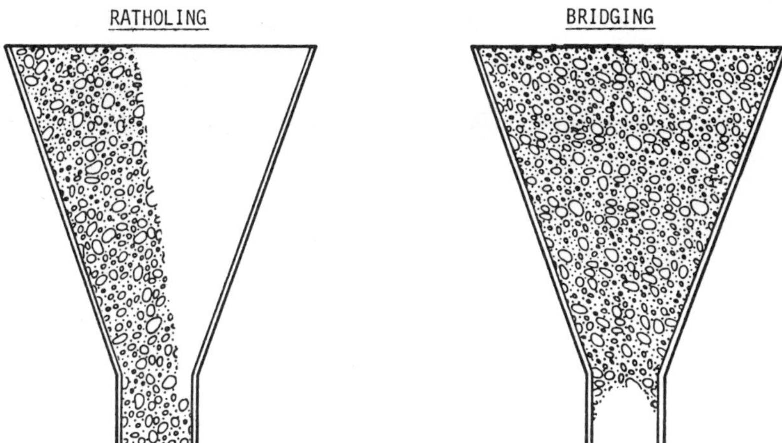

Figure 3-8B. Wood flow handling problems.

Wood fuel as a class has poor flow characteristics, and this must be considered in the silo design. Wood fuel may not be free flowing in silo storage; therefore, silos require active mechanical retrieval systems. These are manufactured in several designs and can include a chain flail or screw auger. Wood fuel in silo storage can bridge over (Figure 3-8B[17]). Under certain conditions, the collapse of this bridge can create outward thrust and/or vacuum conditions severe enough to cause structural failure to the silo.

TABLE 3-4. Steel Silos With Unloader.[22]

STORAGE CAPACITY (FT^3)	WEIGHT OF WOOD FUEL STORED			
	GREEN WOOD CHIPS (TONS)	WOOD PELLETS (TONS)	SILO SIZE (DIA. X HEIGHT: FT)	INSTALLED COST (DOLLARS)
5,000	50 to 60	88	15 x 38	45,000
10,000	100 to 120	175	21 x 33	53,000
15,000	150 to 180	263	21 x 48	61,000
20,000	200 to 240	350	21 x 60	74,000
25,000	250 to 300	438	27 x 50	78,000
30,000	300 to 360	526	27 x 58	90,000
35,000	350 to 420	613	27 x 67	101,000

Notes: 1. Silos are constructed of steel with fused glass.
    2. Height is from floor to top.
    3. Bulk density of wood is assumed to be 23 lbs/ft^3.
    4. Bulk density of pellets is assumed to be 35 lbs/ft^3.

There are a number of successful wood fuel silo storage systems now in operation. Properly designed, they can achieve full automation.

Table 3-4 presents some typical steel silos showing size, storage capacity, and approximate installed costs. Concrete silos are about 12% less on a first cost basis.

IN-PLANT WOOD FUEL HANDLING

This section describes the various methods commonly employed in the transport of wood fuel at a plant site. After delivery, the wood may be transported to long or short term storage, to fuel preparation equipment, or to the boiler hopper. A key design element of wood handling systems is minimization of the amount of transport equipment.

For the nonforest industrial plant, viable transport methods include

- Belt conveyors
- Chain conveyors
- Augers
- Front end loaders

In addition, there is limited application of pneumatic conveying, vibrating conveyors, and bucket elevators.

The demarcation between various transport methods is not always clear, and their recommended uses can overlap. Careful attention is necessary at final design to select the system that will provide good service at reasonable cost with low maintenance. Figure 3-9[16] illustrates four transport methods.

Belt conveyors. Belt conveyors find wide application for conveying wood fuel. Used to convey sawdust, bark, wood chips, and unhogged wood waste, they can handle large capacities and are especially useful for conveying wood fuel over long distances. Belts operate best under uniform loading. They are not satisfactory for retrieving material from storage where the material may be many feet in depth over the belt. Usually belts are limited to an incline of 15° to prevent material slippage. Belts are available with flights and flexible sidewalls which increase the transport angles. Belt conveyors come in various sizes, widths, and lengths—from a few feet up to thousands of feet.

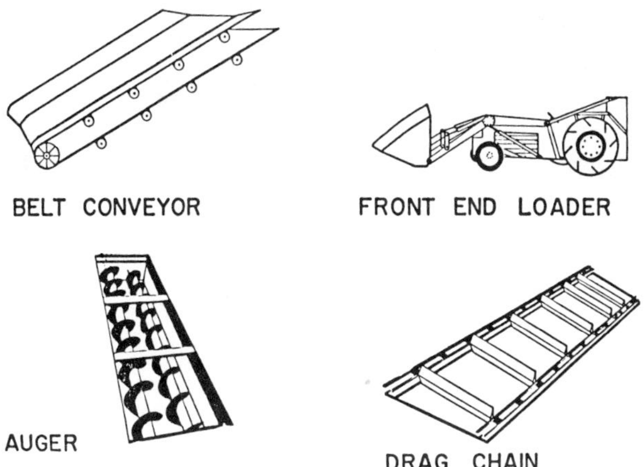

Figure 3-9. Woodyard fuel handling systems.

Drag chains.   Drag chain conveyors are widely used where the transport distance is relatively short. Though construction details differ, they essentially consist of one or more metal chains connected with drag bars running in troughs. Used to transport sawdust, bark, wood chips, and unhogged wood pieces of moderate size, they can be installed in parallel to create live bottoms.

Drag chains are rugged in construction and can handle heavy nonuniform loads. They are especially suited for reclaiming wood from receiving hoppers or storage piles, and for feeding boiler hoppers. They are relatively easy to operate and maintain. Broken links can be replaced or welded by plant personnel.

Augers.   Augers or screw conveyors consist of a helical blade mounted on a shaft enclosed by troughs or covers that hold the material to be transported. They are commonly used in feeders to the burners and as part of the retrieval equipment in silos or bins. They can transport material horizontally or on inclines with reduced capacity. With variable speed drives, they can be used as metering devices. They can handle hogged wood and wood chips. Augers are inexpensive and easy to operate.

Front end loaders.   Front end loaders are useful machines for wood fuel handling. For many installations, the machine can be the principal fuel handling method. For installations that use systems such as drag chains and belt conveyors, the front end loader can clean up spills and serve as backup for equipment breakdown.

There are two basic types—agricultural and construction. These units differ primarily in their design duty. The agricultural tractor is lightweight and is not expected to maintain a heavy duty cycle, whereas construction units often are of articulated design for maneuverability and continuous duty. For wood fuel use, the standard buckets should be replaced with larger volume buckets. Wood fuel at less than 25 lb/ft^3 can be easily handled using high volume buckets—known in the trade as snow buckets or loose material buckets.

TABLE 3-5. Front End Loaders for Wood-Fired Boilers.[22]

TYPE	BUCKET SIZE (YDS^3)	APPLICATIONS BOILER SIZE UP TO	TRACTOR WT (LB)	ESTIMATED FUEL CONSUMPTION (GAL/HR)	ENGINE (HP)	APPROX. COST (DOLLARS)
Agricultural	1½	600 hp or 21,700 lb/hr	6,300	1.0	45	19,000
Agricultural	3	1,000 hp or 34,500 lb/hr	7,800	1.5	64	26,000
Construction	5	3,333 hp or 115,000 lb/hr	21,000	3.5	80	67,000
Construction	12	6,666 hp or 230,000 lb/hr	37,000	7.5	170	150,000

As a size selection aid, Table 3-5 has been prepared. The table is based on the loader being able to move twice the required weight of fuel per hour—that is, the completion of two cycles of scoop, move, dump, and return. Taking into account their more substantial design, a 75% duty cycle is applied to the construction loader while a 25% duty cycle is applied to the agricultural loader.

Other. Among other wood fuel transport methods are pneumatic conveyors, vibrating conveyors, and bucket elevators. Pneumatic conveyors are used in the lumber and furniture manufacturing industries to pick up sawdust for transport to central storage. As operating costs are relatively higher than belt conveying, limited use is expected in the nonforest industry. Vibrating conveyors have certain special applications, but are not widely used. Bucket elevators are modified chain conveyors with buckets attached and are useful where vertical lift is required.

PREPARATION OF WOOD FUEL

The preparation of wood fuel at a nonforest industry plant may be necessary to facilitate handling and to meet the requirements of the burning system. It is normally limited to size reduction and to the removal of foreign objects. In some instances, pre-drying may be appropriate where waste heat is available or where higher flame temperatures are required for processes such as calcining. Many nonforest industries purchase their fuel in a form such that no on-site preparation is required.

Size Reduction

Size reduction is required when the wood fuel has dimensions too large to pass through the burner feed system or to combust properly. For example, two inches may be the maximum dimension desired. If any of the wood fuel exceeds this dimension, it can be reduced to proper sizes by passing through a machine commonly called a "hog" (knife hog or hammermill).

To minimize first cost and operating expense, on-site preparation of wood fuel should be kept to the minimum consistent with the needs of the combustion system. Therefore, prior to the purchase of a particular unit, the availability of suitable wood fuel should be evaluated. For example, some wood burning units require dry or partially dried wood. If only green fuel wood is available locally at reasonable cost, then only systems that are capable of using green wood should be considered.

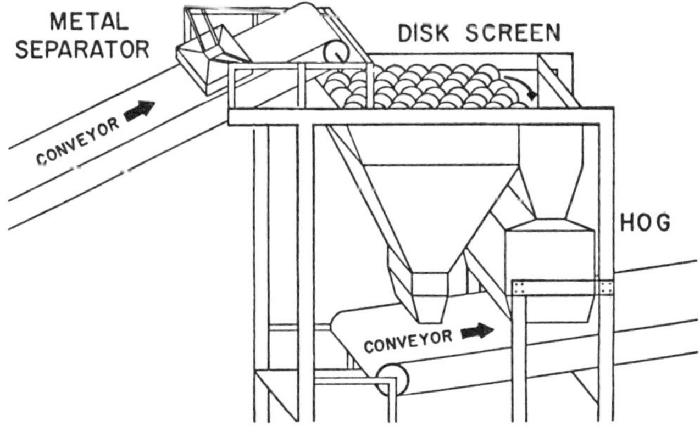

Figure 3-10. Wood fuel size reduction system.

Figure 3-10[22] illustrates the arrangement of a typical size reduction system utilizing waste wood. The unhogged material first passes under a magnetic metal separator which removes tramp iron or steel. (There are several designs of magnetic separation devices, but all work on the principal of the magnetic attraction of iron.) The material then passes onto the top of the disk screen. The disk screen consists of a series of disks rotating on parallel shafts. The spacing of the disks is such that material over a certain size (e.g., 2 in.) will pass across the top of the disk screen and drop into the hog. The hog will then reduce the larger pieces to the required size (e.g., 2 in.). The purpose of the screen is to reduce the horsepower requirements of the hog by screening out the material that is of proper size. In the usual design, the material passing through the screen and the hog is forwarded onto a conveyor and transported to storage. Costs of some typical screens, hogs, and hammermills are listed in Table 3-6.

TABLE 3-6. Wood Fuel Size Reduction Equipment: Costs.

	CAPACITY (TONS/HR)	COST (DOLLARS)
Screen Costs [22]		
Small Disc	40	24,000
Large Disc	250	60,000
Small Shaker	10	6,500
Large Shaker	40	24,500
Revolving Drum	40	54,000
Costs of Hogs and Hammermills		
Small Knife Hog	10	14,500
Large Knife Hog	32	45,000
Small Hammermill	3	12,000
Large Hammermill	8	29,000

WOOD FUEL DRYING

The net effect of pre-drying on the combustion of wood is that dry wood is more easily burned and flame temperatures of 2300 to 2500°F are obtainable (instead of 1800°F possible with green wood). Also, the overall thermal efficiency of a boiler can be increased (5 to 15%).

Drying Methods: Wood Piles

Research done by Mssrs. White and DeLuca (of V.P.I.) several years ago indicates that the elevated temperatures associated with bulk storage wood piles result in significant moisture content reduction, coupled with a probable net decrease in pile heat content. Piles of different geometries were investigated. The piles were variously mixed green sawdust, hardwood bark, and pine bark. All piles exhibited a rapid increase in internal pile temperature—with the greatest temperature near the geometric center at the base of the pile. The temperatures rapidly reached a threshold value and thereafter remained fairly steady, varying from steady state levels of 120 to 170°F. Although some decrease in moisture content was observed, this research is not conclusive. Research on this subject—especially regarding whole tree chips—is ongoing here at Georgia Tech, by the Georgia Forestry Service, and by Professor Mark White. More definitive results in this area are expected. Gains in net heating content due to moisture content decrease during open pile storage periods are tempered, and may well be offset, by saturation of the surface layer by precipitation and evaporation of volatiles. Piles with large surface volume/total volume ratios (i.e. "small" piles) are particularly sensitive to surface saturation from precipitation (see Figures 3-3, 3-5, 3-6).

Wood Drying Equipment

A wide range of gaseous, liquid, and solid fuel dryers have long been used in many industries. Single-pass and triple-pass fossil fuel fired, motor driven, rotary dryers are commonly used in fertilizer plants, grain, and other process drying applications. Cascade dryers and flash-type dryers are in widespread use in many processes. With the spiraling cost of energy, equipment vendors and biomass users have, in the last decade, adapted these existing technologies to drying wood. In applications where exact moisture contents are not required by the process and where the existing boiler has solid-fuel firing capability and is presently firing fossil fuel, waste heat derived from flue gas is a demonstrated economical drying agent with short paybacks and good returns on investment. Paybacks on retrofitted waste-heat dryers for systems already burning (high moisture content) wood will tend to be not so dramatic as those calculated for displacement of fossil fuel with dry wood fuel.

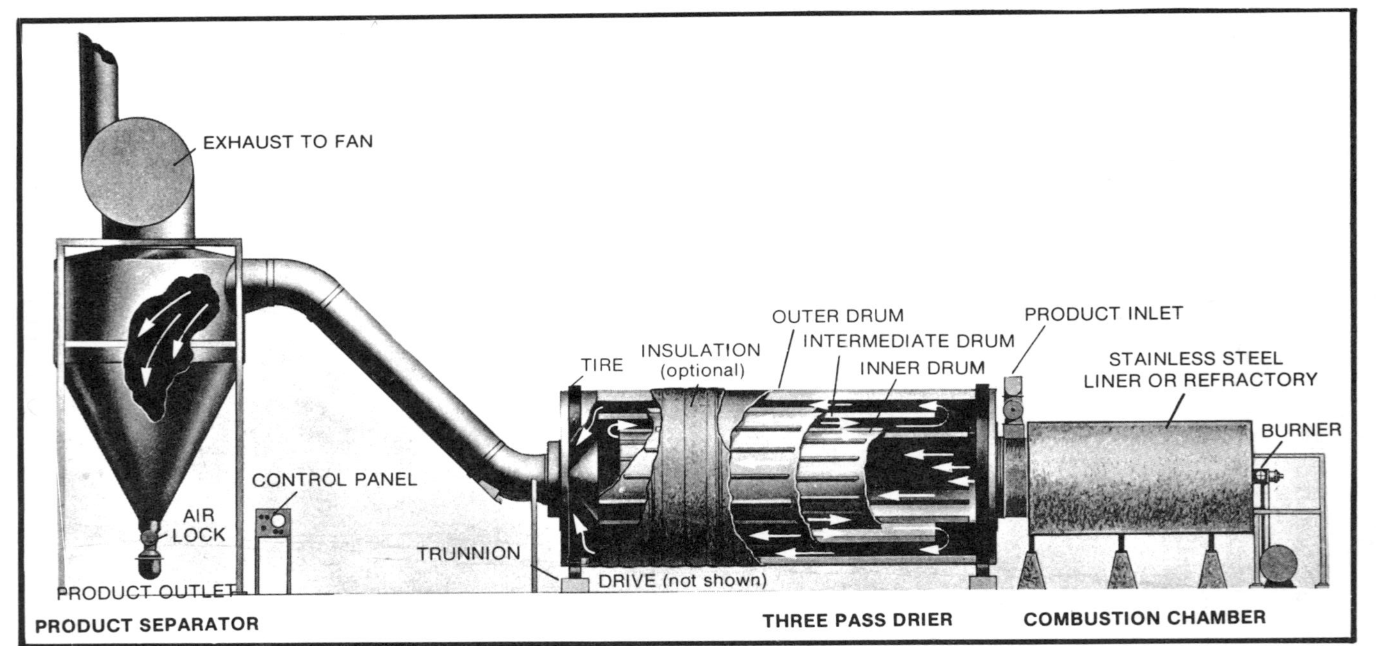

Figure 3-11. Rotary wood drying system. (*courtesy Aeroglide Corporation.*)

Rotary Dryers.   Rotary dryers are of two types—single pass and triple pass. The single-pass dryer requires a smaller pressure drop, and consumes less fan horsepower. Consequently, it has a lower operating cost. Control of product moisture content is not so precise as in a triple-pass dryer; however, this should not be a problem where the end goal of the dryer output is combustion. Both types are available from well known manufacturers. Koppers, Manufacturing Engineering Construction (MEC), Rader, and Aeroglide are among the many vendors with waste-heat fired single- and triple-pass dryers (Figure 3-11). A waste-heat rotary dryer is operating in Augusta, Georgia, on a wood-fired brick kiln. A single-pass rotary dryer will typically process 6 tons/hr of 50% moisture content mill residue in the waste-heat drying mode.

Cascade Dryers and Flash Dryers.   In cascade dryers (Figure 3-12[20]), wood fuel is dried by falling through streams of hot gas—much like the path of lightweight sand would appear when tossed into a cascading fountain. Cascade dryers tend to be large capacity. No auxiliary fuel input or motor horsepower for rotation are required. However, the pressure drop demand across the unit must be taken up by the fan, so some energy input is necessary. Flash-type dryers are simply several loops of duct where the wet material and hot flue gas mingle and drying occurs. Successful cascade dryers have been reported from Sweden. The first American installation of a Swedish-designed cascade dryer was completed at a paper mill in upstate New York in mid-1980. The power plant superintendent reports he is satisfied with its operation even as fine tuning of controls and operator education proceeds. A very sophisticated flash drying system, incorporating a ¼ in. minus hogger in the middle of the drying loop, is being tested on two full size installations in Sweden; however, the vendor has not placed the unit on the market. In addition, the vendor has added a pellet mill to the equipment train. The pelletizing capability enables the mill to store dry, pelletized wood which is ready for stoking or pulverized firing. (At this time, plant decisions involving pelletization must recognize the high equipment and operating costs inherent thereto).

All drying equipment described requires similar auxiliary equipment for material collection and fly ash removal (Figure 3-13). Systems of single cyclones, multiclones, and, where necessary, baghouses, scrubbers, and precipitators are common. Usually larger wood particles are collected in a precollection hopper while bark fines are separated in

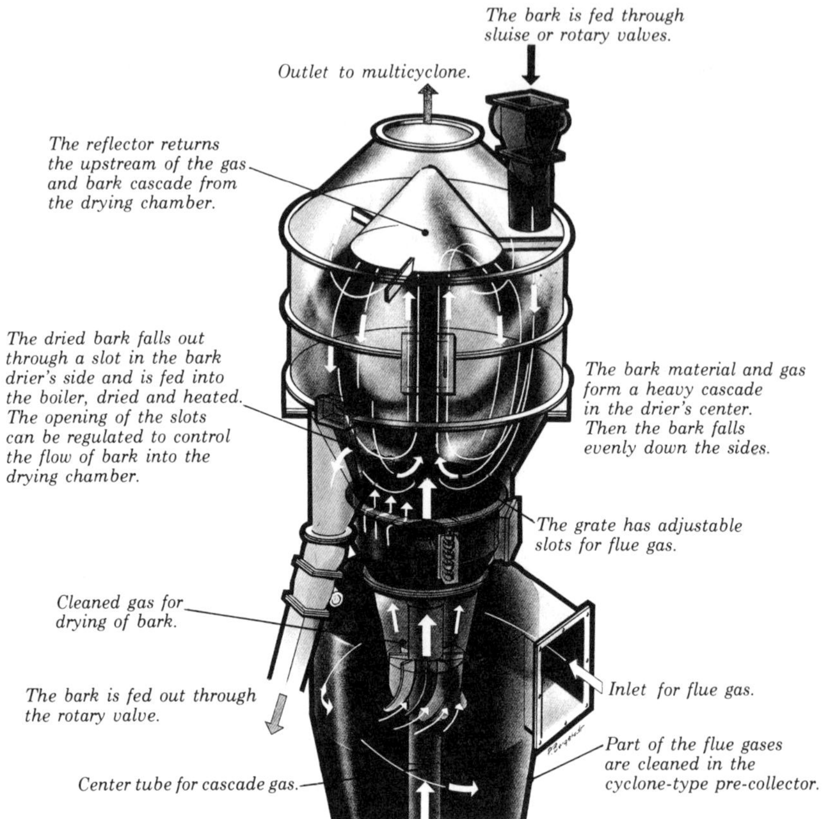

Figure 3-12. Cascade bark drying system. (*courtesy Bahco Systems Inc.*)

cyclones. The dried fuel is conveyed to a storage bin and then into a standard fuel feed system. Combustibles from the fly ash are collected and mixed with the bark fuel while the remaining fly ash is removed by the pollution control equipment.

APPROPRIATE APPLICATIONS OF WOOD FUEL DRYERS FOR NONFOREST PRODUCTS INDUSTRIES

Retrofits

Retrofitting waste-heat wood fuel dryers to existing solid fuel boilers represents the largest market for drying equipment. The reasons for this

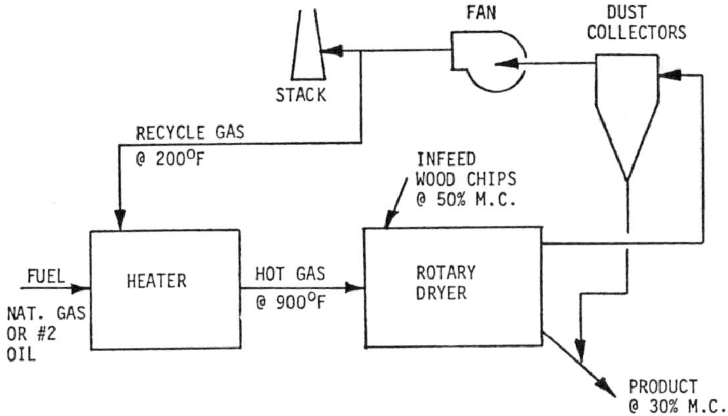

Figure 3-13. Wood drying system flow chart. (*courtesy Stearns-Rogers Manufacturers, Inc.*)

are economic. Boiler efficiency improvement is effected by decreasing the temperature of exiting flue gas from the neighborhood of 450 to about 200°F.* Flame temperature increases, less air is moved through the boiler by the fans, and ease of flue gas particulate collection improves due to more complete burning of the fuel in the combustion zone. Often, a plant with steam requirements greater than current operating capacity can gain enough capacity from burning lower moisture content fuel that installation of additional peak-load steam generating equipment can be avoided. Also, hogged wood or sawdust of greater than 65% moisture content will not normally sustain combustion. In the past, some plants have used fossil-fuel fired dryers to dry this material so as to avoid storage and/or disposal problems. The advent of dryers fired by waste heat make this wet fuel a potentially attractive fuel source.

In processes where there are demands for high temperature off-gases (kilns, textile drying, etc.) increased flame temperatures can be achieved by using dry fuel. The brick industries of Georgia and South Carolina, active in conversion to wood fuels (from natural gas) for firing kilns, have been successful in utilizing waste-heat fired rotary dryers for obtaining the dry sawdust required for their process.

---

*Due to the SO_2 acid dewpoint problem, flue gas exit temperatures of 200°F are feasible only where no sulfur-bearing fossil fuels are being combusted. Where fossil fuel is being combusted in conjunction with wood, flue gas exit temperatures must be maintained at greater than approximately 275°F—the approximate SO_x acid dewpoint. Recently, indications from the field are that allowing exiting (wood-derived) flue gas to pass below 350°F permits undesirable formations on ducts, fans, etc.

New Installations

Dryer manufacturers claim the following benefits are gained when a flue gas dryer is included in the scope of new power plant construction:

a. reduced dimensions of boiler heat transfer surfaces, air preheaters, and economizers
b. higher efficiency
c. improved load change response
d. overall lower costs for new plants incorporating waste-heat wood dryers

Feasibility for Boiler Steam Plants

Existing solid fuel boilers of at least 60,000 lbs/hr steam capacity are generally the least size boilers that should be considered for dryer retrofits. Of course, the larger the boiler and the more expensive the fossil fuel, the more attractive investment returns become. Installed costs for dryers are somewhat site specific. A very rough rule-of-thumb for installed cost is $38,000 per ton/hr of moist fuel input. Payback periods range from one to three years. Dryer installations are eligible for the 10% equipment investment benefit and may qualify for the 10% energy investment credit.

The growth of the wood densification (pelletization) industry represents an area where waste-heat wood dryers should be of great value. At present, some pellet mills (still in the demonstration phase) are using conventional fossil-fuel fired rotary dryers to sufficiently dry wood in preparation for pelletization.

TYPICAL DESIGNS OF WOOD FUEL FACILITIES

The purpose of this section is to present three conceptual designs for wood fuel facilities that address the needs of nonforest-related industrial plants. The designs can be adapted to meet the requirements for any size plant. Bear in mind that actual designs are site specific and should account for local conditions.

The first system is based on a silo (for storage) and uses prepared fuel (fuel which requires no on-site beneficiation). The second system is based on prepared fuel using both outdoor and covered storage. The third sys-

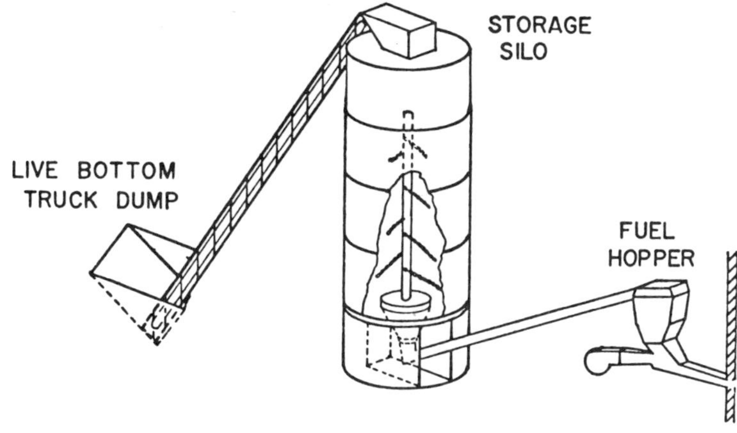

Figure 3-14. Silo wood handling system.

tem is based on wood waste material and contains a fuel preparation section—utilizing open and covered storage.

The silo-based system (Figure 3-14[22]) is a commonly used design for the moderate size plant and for locations where storage area is limited. This system is also indicated where the wood fuel is dry and protection is needed from precipitation. The silo can be made of steel, poured concrete, or staved concrete construction and may be almost completely automated. It requires little attention or maintenance.

Silos are subject to fuel flow interruptions from bridging of the fuel in the silo. For this reason, most silos include a device for breaking up the bridged fuel to assure free flow. There are several devices of this kind. The drawing indicates a frequently used unit which consists of a motor-driven vertical shaft with chain flails. The flow problem also limits the size of the material. Large pieces will not pass through the silo and conveying system, so the wood fuel is limited to pieces 2 in. and smaller. These systems are often designed for three days' storage. With this limited storage, a reliable fuel supply is imperative.

The next system (Figure 3-15[22]) consists of open and covered storage with wood handling done by a front end loader. For a given size, this design represents the lowest capital cost. No conveying equipment is specified, though for a particular installation, conveying equipment may be desirable.

TYPICAL STORAGE LAYOUT

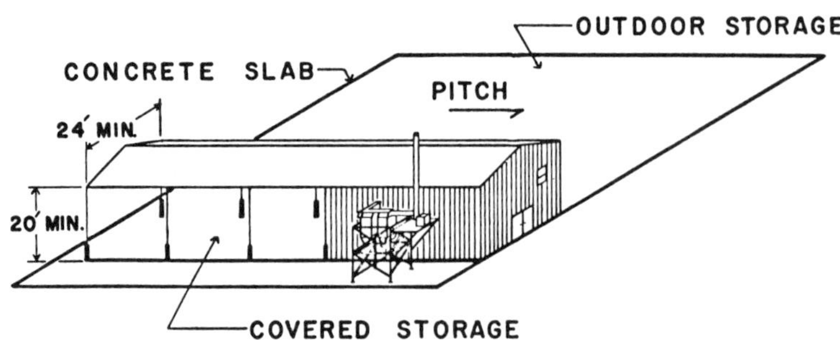

Figure 3-15. Outdoor and covered storage.

The open storage consists of a concrete slab designed to hold a 30-day supply of wood fuel. The slab has a slight pitch to aid in rainwater runoff. The covered storage is designed to hold three days of fuel and consists of a concrete slab with an open-sided metal building. In practice, the covered storage is the active area with the outdoor storage

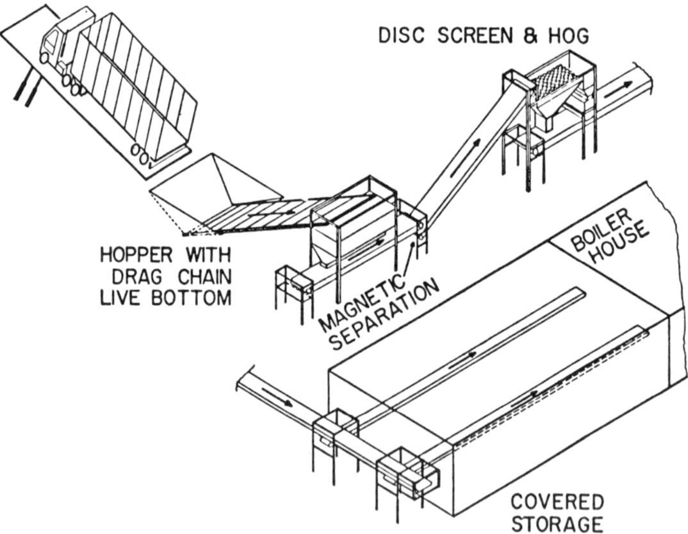

Figure 3-16. Large woodyard fuel handling systems.

serving as a reserve holding area to allow for special purchases and to provide fuel during periods when the regular supply may be interrupted.

No fuel preparation is included, as it is anticipated that the fuel will be purchased sized to the requirements of the combustion system. A truck receiving-unloading system may have to be included depending upon the type of fuel delivery.

The last system (Figure 3-16[22]) illustrates the components that are usually necessary for the larger systems. The fuel is received from a hydraulic truck dump into a live bottom pit. The material is then conveyed up through a station where it is passed over a magnetic separation belt to remove tramp metal. Next, the fuel is conveyed to a disc screen and hog where any oversized pieces are reduced to proper size. From the hog, the material is transported by conveyors to the covered storage. (Not shown but generally included in a system of this size is an open storage area.) After passing through the disc screen and hog, the conveying system is so arranged that the fuel can be transported to covered storage or to an open storage area. A front end loader is a necessary piece of equipment to handle special transport problems and to handle spills.

4
Cogeneration

INTRODUCTION

Cogeneration can be defined as the simultaneous production of electricity and thermal energy. In the case of a wood boiler, the thermal energy will be in the form of process steam. An industrial plant planning a major expansion of steam producing capabilities would do well to examine the cogeneration option.

Hardware is available to the industrial energy user for generation of his own power, and the source of energy for the steam turbines could well be a wood-fired boiler system. The mechanics of working out cooperative agreements with local power companies are not clearly defined at this time, although some utilities and REA organizations are becoming more receptive to the idea of buying back power from small plants. Most large pulp and paper mills in the Southeast generate all or part of their electricity (largely supplied by wood-fired boilers) and some have sold power to utility grids, but this situation is not widespread. Utility companies usually cite these major drawbacks: (1) industrially generated power is often of low quality (inadequate frequency and voltage control) and (2) the industrial producer often produces excess power at off-peak times of the day when it is not needed by the utility.

GOVERNMENT INCENTIVES

An impediment to the wide acceptance of cogeneration has been the unanswered question of cogenerator/utility interface. Often potential cogenerators were discouraged because utilities were under no obligation to buy excess power. When electricity was offered for sale, the cogenerators could then be subject to the same strict laws as a utility. In

March 1980, however, federal legislation was enacted to encourage further cogeneration ventures.

These rules state that utilities must purchase electricity from cogenerators at the utilities' "avoided" cost. An avoided cost is that amount a utility saves in fuel and capacity additions costs. Legislation favoring cogeneration could influence many suitable industries to consider this alternative.

STEAM TURBINES

The prime mover used for cogeneration of process steam and electricity is a steam turbine. Steam turbines can be grouped very broadly into condensing or non-condensing units. Since steam exhaust is condensed at the exhaust of a condensing turbine, process steam at the desired pressure must be extracted before the exhaust. Extraction valves are then included to supply the process steam. In the non-condensing case, the turbine produces power by acting as a pressure reducer and the turbine exhaust becomes process steam. (TOPPING CYCLE)

The selection of a condensing or non-condensing turbine is based on such factors as boiler pressure, process steam requirements, electrical requirements, and cooling water availability. Steam turbine types suitable for industrial cogeneration are listed in Table 4-1. If the most important boiler product is process steam, generally a non-condensing turbine is specified. Operation requires increasing the boiler pressure which is used in the turbine to produce power. In the non-condensing unit, process steam pressure and flow are requirements to be satisfied. Therefore, since electric generation is a direct function of steam flow, periods

TABLE 4-1. Turbine-Generator Configurations.[17]

TURBINE TYPE	APPLICATION
Back Pressure (Non-Condensing)	Turbine exhaust meets process steam demand, electricity generation directly related to steam flow
Single Automatic Extraction (Condensing)	Supplies process steam at one pressure level, meets variations in electrical load
Single Automatic Extraction (Non-Condensing)	Supplies process steam at two pressure levels, electrical generation directly related to steam flow
Double Automatic Extraction (Condensing)	Supplies process steam at two pressure levels, meets variations in electrical load
Double Automatic Extraction (Non-Condensing)	Supplies process steam at three pressure levels, electrical generation directly related to steam flow

of low steam demand correspond to low electrical generation. Non-condensing turbines cannot meet all variations in electrical load and an alternate power source must be provided.

If the turbine-generator set is expected to meet all variations in both process steam and electrical load, a condensing turbine is necessary. Power is generated by steam flowing through the turbine and process steam is provided by extraction. In times of high steam demand, most of the power is generated by the steam before it is extracted. At other times, the entire steam flow is through the turbine to the condenser and electricity is produced. This arrangement allows variation of process steam and electrical production independently. Three disadvantages of condensing turbines are system size limitations, high cost, and overall cycle efficiency. Condensing turbines are difficult to produce in the smaller power ranges and sizes below 5,000 Kw are not available. Overall system cost is increased with the addition of a condenser, piping, and a circulating water system. Overall cycle efficiency with a condensing unit is reduced as virtually all the usable heat contained in the exhaust steam is lost to the cooling medium in the condenser.

Manufacturers were surveyed to get representative cost figures for four different sized cogeneration systems. Results for 1, 5, 10, and 25 Mw installations are presented in Table 4-2. Installations include turbine, generator, baseplate, and lube systems. Piping, switchgear, condenser, and other auxiliary elements are not included. Although there are variations, some general conclusions can be reached. For a one megawatt installation, the turbine suitable is a single stage non-condensing unit. In the power ranges above this, either a condensing or non-condensing unit can be used. Most manufacturers specified a non-condensing unit, although one was quoted both ways. Turbine maintenance costs varied widely, but all were calculated one of two ways. Annual maintenance costs are either estimated as a percentage of capital investment (2 to 2½%) or as a multiplier times power production (.3 to .4¢/Kwh).

ECONOMIC CONSIDERATIONS

The alternatives to be considered when deciding upon a cogeneration system are:

- Purchase power with separate steam generation
- Cogeneration of power and steam
  —Back pressure turbine
  —Condensing turbine

TABLE 4-2. Turbine Price Estimates.[17]

MANUFACTURER	A	B	C	D	E	COMMENTS
1 Mw	$ 500,000	$ 125,000	$ 300,000	$130,000	$ 150,000	Single Stage
	(500 $/kw)	(125 $/kw)	(300 $/kw)	(130 $/kw)	(150 $/kw)	Non-Condensing
Maint-Annual	$ 12,500	$ 7,000				
5 Mw	$1,180,000	$ 770,000	$ 750,000	$600,000	$ 600,000	Multi-Stage Condensing
	(236 $/kw)	(154 $/kw)	(150 $/kw)	(120 $/kw)	(120 $/kw)	or Non-Condensing
Maint-Annual	$ 29,500	$ 35,000		$ 12,000		
10 Mw	$1,570,000	$1,200,000	$1,150,000		$1,100,000	Multi-Stage Condensing
	(157 $/kw)	(120 $/kw)	(115 $/kw)			or Non-Condensing
Maint-Annual	$ 39,500	$ 60,000				
25 Mw	$3,450,000	$2,500,000			$3,400,000	Multi-Stage Condensing
	(138 $/kw)	(100 $/kw)				or Non-Condensing
Maint-Annual	$ 86,000	$ 190,000				

Factors which influence the decision include purchased electrical cost, fuel cost, and required investment ($/kw). General cost figures for turbine generator equipment are shown in Table 4-3. Note the increased cost for the condensing turbine option. As an illustration, consider a plant with the following requirements:

- Steam flow   101,500 lb/hr
- Steam pressure   250 psig
- Steam temperature   445°F

TABLE 4-3. Turbine Costs: Condensing vs. Non-Condensing[17]

	NON-CONDENSING	CONDENSING
Turbine-Generator	$360,000	$700,000
Condenser	—	100,000
Cooling Tower	—	90,000
	$360,000 ($144/kw)	$890,000 ($178/kw)

NOTE: Does not include piping, insulation, instrumentation, electrical or switchgear

Figure 4-1[17] is a typical performance map for a steam turbine. This performance map is used to illustrate the output of the turbine supplying process steam at the conditions given above.

The line A-F-B shows all the possible operating conditions for the back pressure case. Electrical output is directly proportional to steam flow, and turbine rated output is reached at rated process steam flow. This line would also be the operating line for the condensing case when all the steam is being extracted and none flows to the condenser. Using the condenser for control allows the condensing turbine to operate anywhere in the area bounded by A-B-E-C. The performance map illustrates that power is produced both by the extraction flow and flow in the condenser. To supply the turbine, the boiler pressure is increased to 650 psig in both the condensing and noncondensing cases, but the steam flow is increased only for the condensing case.

Figures 4-2 through 4-6[17] show the operating conditions for the boiler alone and with four different turbine options. The heat balances for each case are shown in Figures 4-3 through 4-6. The boiler only case shows fuel consumption, increased fuel consumption over the boiler only case,

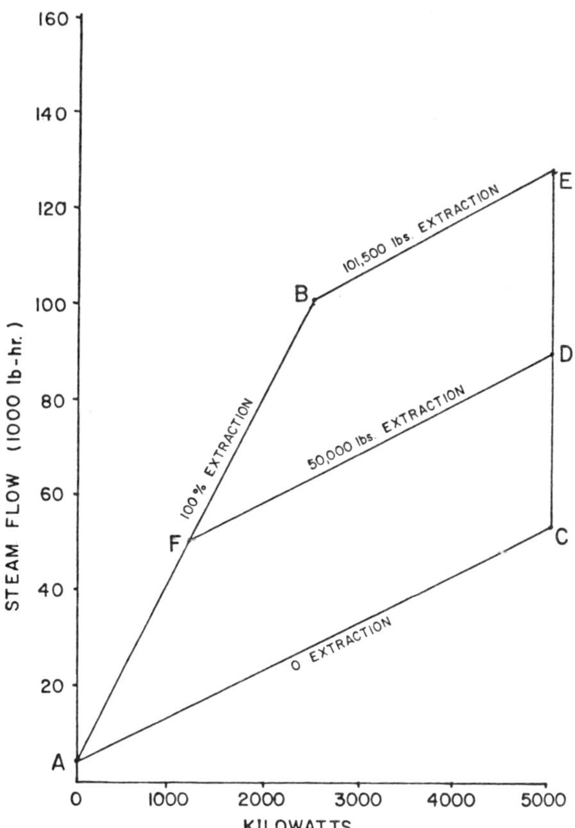

Figure 4-1. Steam turbine performance map.

and the electrical cost using fuel at $1.65/MMBtu ($15/ton) for four turbine cases. For the back pressure case, electricity is generated by steam required for the process. Two extraction cases are examined: (1) Case A in which half the power is generated by process steam before extraction and half is generated by steam flow to the condensers; and (2) Case B where one-third is generated by extraction steam and two-thirds by condenser steam flow. The final case, that of the condensing turbine, is shown as a limiting case. Notice that electrical costs (Table 4-4) are higher for extraction turbines than for back pressure turbines. The electrical cost increases in proportion to the amount of generation by condenser flow and reaches the limit in the condensing only turbine. The difference in electrical cost is due to the relative inefficiency of con-

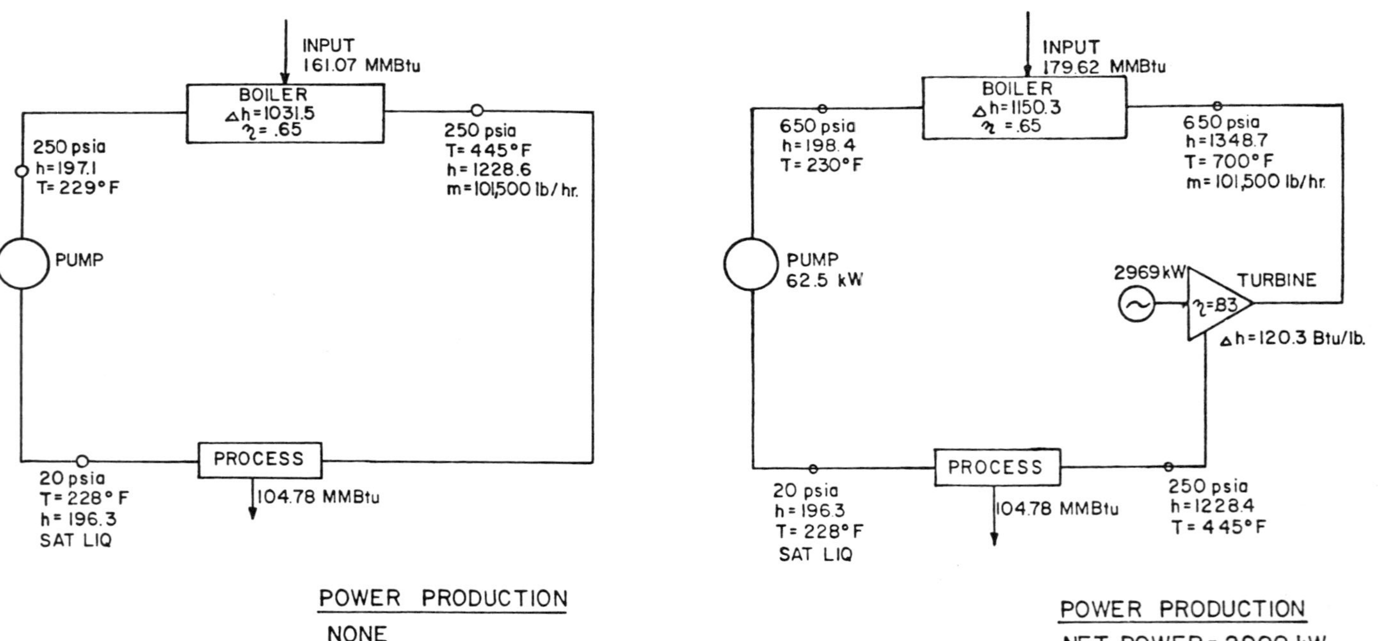

Figure 4-2. Wood-fired boiler cycle.

Figure 4-3. Back pressure cycle (non-condensing turbine).

Figure 4-4. Extraction turbine cycle (case A).

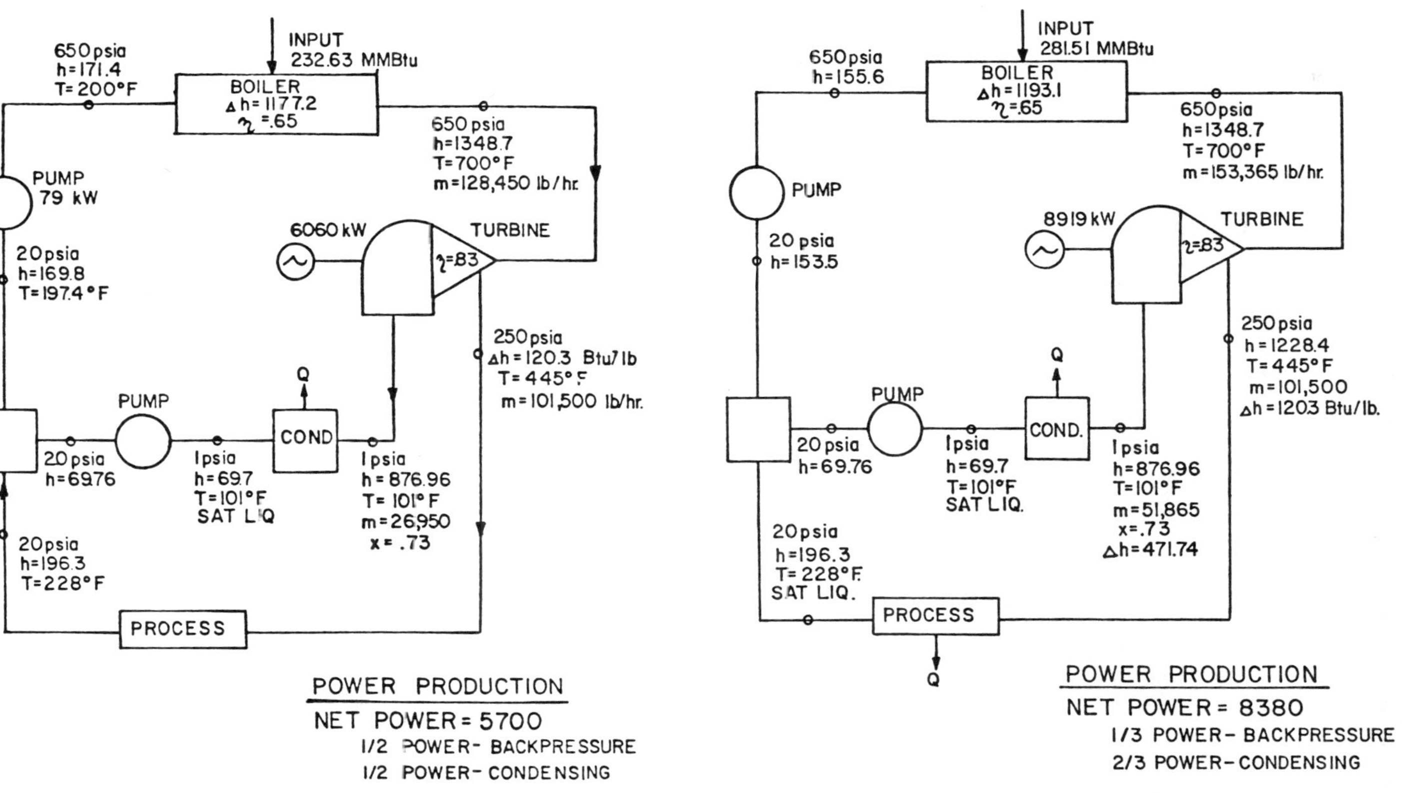

Figure 4-5. Extraction turbine cycle (case B).

TABLE 4-4. Turbine Operating Conditions and Electrical Production Costs.[17]

	BOILER ONLY	BACK PRESSURE	EXTRACTION(A)	EXTRACTION(B)	CONDENSING
Boiler Pressure (psig)	250.00	650.00	650.00	650.00	650.00
Boiler Temperature (°F)	445.00	700.00	700.00	700.00	700.00
Steam Flow (lb/hr)	101,500.00	101,500.00	128,450.00	153,365.00	43,850.00
Electrical Output (kw)	—	2,900.00	5,700.00	8,380.00	5,700.00
Fuel Consumption (MMBtu/hr)	161.07	179.62	232.63	281.51	86.12
Increased Consumption (MMBtu/hr)	0	18.55	71.56	120.44	—
Electrical Cost (¢/kw)	—	1.05	2.07	2.37	2.49

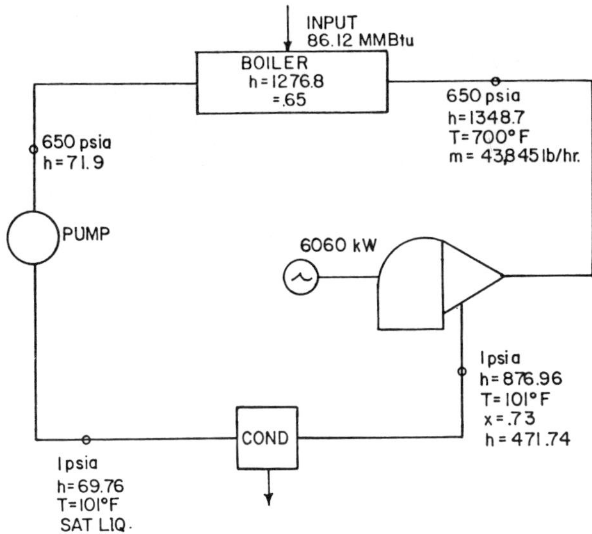

Figure 4-6. Condensing cycle (no extraction).

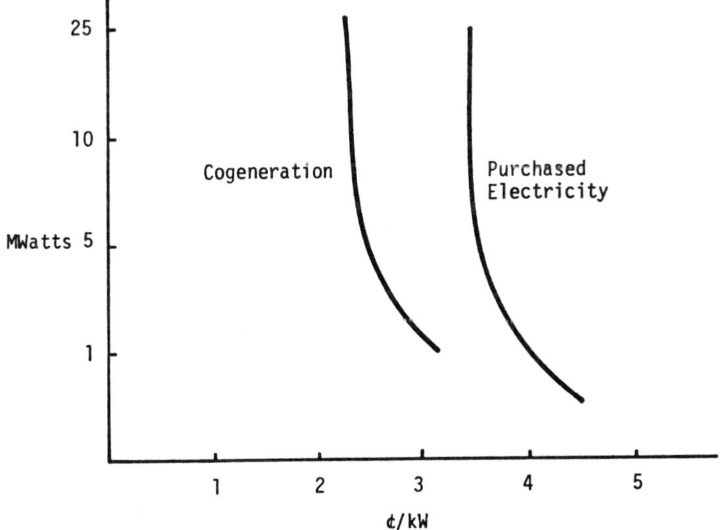

Figure 4-7. Power costs comparison.

densing versus back pressure turbines in this application and reflects the cost of control being provided by the condenser.

Although unit operating costs are less for the back pressure turbine, other considerations must play a role in the selection. The extraction turbine costs more to install, yet power production is greater. Also, the extraction turbine provides power on a consistent basis while back pressure turbine output varies with process steam flow. Thus, when using a back pressure turbine, a reliable outside source of power must be available. Due to the wide surges in output from a back pressure turbine, the additional kw requirements can seldom be purchased at a low rate.

Figure 4-7 compares costs for cogenerated versus purchased power. Notice that the cogeneration curve ends at about 1 megawatt. The important conclusion to be gained from these curves is that for the plant considering purchasing a new boiler in the 30,000 to 50,000 lb-steam/hr range the additional cost of cogeneration equipment becomes highly justifiable.

COGENERATION SYSTEM SIZE

Cogeneration systems in large sizes (above 20 Mw) have been in wide use for many years; however, rising utility costs have resulted in many smaller scale users examining this option. Several disadvantages tend not to favor electrical production from boilers smaller than 50,000 lb/hr. The efficiency is low for this size of steam turbine, with steam rates as high as 60 lb/kw hr meaning decreased electrical production. Table 4-5 shows the expected power production from a back pressure turbine for four different steam flow rates. Also, because of their size, no economy of scale is realized with steam turbines this small and capital cost can be roughly three times as much per kilowatt as large steam turbines. The electrical production can be increased if a condensing unit is uti-

TABLE 4-5. Electrical Production for Various Steam Flow Rates.

STEAM FLOW (LB/HR)	APPROXIMATE STEAM RATE (KW/LB/HR)	ELECTRICAL PRODUCTION (KW)
10,000	55.50	180
25,000	46.30	540
50,000	41.67	1,200
100,000	40.00	2,500

lized. Use of a condensing turbine means that little or no steam will be available for process applications, and this is unacceptable for most industrial situations since process steam is the primary purpose for having a boiler. In summary, electrical generators for cogeneration systems can be purchased in all sizes; however, economics tend to favor turbines of 5 Mw and larger.

5

Wood-fired Boiler Emissions and Control

INTRODUCTION

The primary environmental cause for concern in the burning of wood is the emission of particulate matter resulting from the combustion process. The presence of smoke indicates the presence of particulates, but the relationship between smoke and particulates is not easily quantified. Smoke results primarily from a combination of inorganic ash, non-combustibles, and particles of carbon and other combustible matter that have not burned completely. For wood combustion, the emission of most concern is particulate matter related to the ash content of wood. A dark plume from an industrial stack can indicate poor maintenance and poor operational practice, or an unavoidable situation such as a rapid load change or boiler upset.

The 1976 Clean Air Act established or caused to be established rules and regulations that govern the emission of certain materials into the atmosphere. As it is an expense that can be substantial—both in initial and operating costs—emission control deserves careful attention from the system designer.

Emission control devices cannot be considered in isolation from the other parts of the system. Fuel, burner, and boiler design all influence the particulate emission rate. For example, with low turbulence combustion, more of the ash remains in the combustion chamber and does not pass through the boiler. The emphasis must be on system design to provide an installation that will comply with regulation.

TYPES OF EMISSION

Particulate from the combustion of wood is composed of ash, unburned char, condensed droplets, sand, and other silicates foreign to the fuel itself. As do other combustion sources, wood furnaces emit carbon monoxide (CO), oxides of nitrogen (NO_x), trace amounts of oxides of sulfur (SO_x), and unburned hydrocarbons (HC); however, these emissions are either too low to be of concern or are not regulated by law. Because wood is lower in ash content than coal, it tends to produce less particulate and almost no sulfur oxides when compared with coal. From an emissions viewpoint, wood can be an attractive alternative.

EMISSION REGULATIONS

Applicable air pollution standards for emissions vary from state to state. In Georgia, the emission criteria to be considered by prospective operators of wood-fired boilers is included in "Rules and Regulations of Air Quality Control" published by the Georgia Department of Natural Resources. Figure 5-1[10] illustrates the maximum permissible particulate emissions from fuel burning installations. Allowable emissions are based on total input in millions of Btu's per hour, beginning on the horizontal scale with 1 million Btu's per hour. Varying from state to state, the applicable regulations should be determined for each location.

"Percent plume opacity" refers to the amount of sunlight blocked by

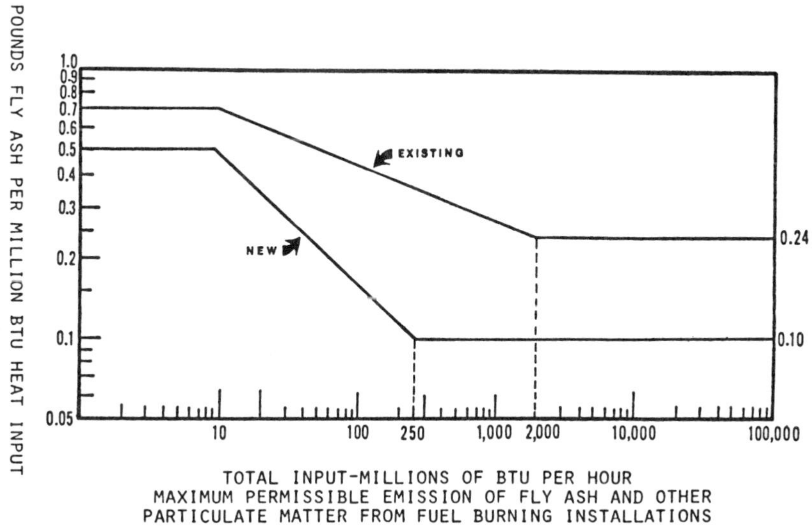

Figure 5-1. *Maximum Permissible Emissions of Flyash and Particulate* (State of Georgia).

the source plume. An opacity of 100% will theoretically allow no light to pass through, while the background will only be obscured 20% by a plume of 20% opacity. The State of Georgia recently enacted legislation making the general opacity limit 20% or less except for 6 minute-periods within a given hour when 27% opacity is allowed for two minutes to permit soot blowing or system upsets. Wood-fired units of moderate to large size will be hard pressed to meet opacity standards and the accompanying particulate regulation without some form of control equipment. Completion of smoke reading schools conducted by the Georgia Department of Natural Resources (DNR)—three-day sessions held several times per year—will result in the attendee becoming a Certified Smoke Reader. This experience is worthwhile for any plant engineer, if only to gain knowledge of regulatory agency procedures.

Companies buying a boiler without help from a design firm should be aware of the appropriate emission control standards. Before constructing a wood-fired boiler in Georgia, one should submit form APCS-APC-2 (an application to construct fuel burning equipment) to the Air Quality Control section of the Department of Natural Resources. This form is reviewed, and if in the agency's judgment the source will comply with existing emissions control standards, it is approved.

The critical path on which the startup of new combustion equipment travels often involves crossing a bridge of lengthy paperwork. For this reason, it is recommended that air pollution regulatory authorities be involved in the project from the earliest possible stage, thereby minimizing risks of expensive and time consuming installation redesign.

There is another regulation which may apply depending on the outcome of a review. To maintain air quality at the current levels, the U.S. EPA promulgated regulations (to prevent significant deterioration (PSD) of air) for sources that emit above 250 tpy (tons per year) of wood fly ash. Generally, for the majority of small industrial boilers, PSD regulations will not apply. All sources not regulated by PSD standards will fall under the existing state particulate and opacity rules already presented. An EPA workshop manual, "Prevention of Significant Deterioration" (October 1980), is available in conjunction with workshop sessions for interested parties (see Bibliography Ref. No. 52).

After construction, the owner must file for a permit to operate. If no significant design changes have been made since filing for a permit to construct, a short form to operate may be sufficient. If significant changes have been made, another form APCS-APC-2 may be necessary.

EXPECTED EMISSIONS

Areas Affecting Emissions

Combustion and the formation of emissions is a complicated process controlled by numerous variables. The three major areas that affect wood-fired boiler emissions are fuel, boiler design, and boiler operation.

Fuel.　The importance of fuel size and size consistency cannot be over-emphasized. Oversize pieces are difficult to distribute evenly in the furnace and tend to burn slowly. When suspension burning occurs, fine particles are easily entrained in the flue gases—leading to higher particulate loading. The size of the fuel should be consistent with equipment design parameters, aiding distribution on the bed and assuring equitable distribution of the underfire air. The cleanliness of the fuel must be considered in terms of whatever foreign matter is picked up during harvest and transport and then fed into the boiler. Ash content of wood itself is low; but material lodged in the bark during removal from the forest, or scooped up by the loader during handling, effectively increases the non-combustible fraction of the woodfuel.

Boiler design.　Wood-burning boilers have design specifications that tend to increase the efficiency of combustion and limit emissions. Designers strive to minimize the gas velocity in the boiler by increasing the grate area. A large grate area decreases the air flow per unit resulting in lower velocities. Lower velocities help reduce emissions by increasing the fuel residence time in the combustion zone and decreasing entrainment of fine particles by high speed air. Another factor that affects emissions is the method by which fuel is introduced into the boiler. Spreader stokers will have some suspension burning and generally higher emissions than methods that burn primarily on the grate. Particles burning in suspension are easily carried out of the boiler. It is a common experience with wood burning that a percentage of the fuel may not completely combust and is carried out of the boiler unburned, to be caught in the primary collector. Especially in large size boilers, the unburned carbon can be separated from the fly ash by a pre-collection hopper or a gravity separation device and re-injected into the boiler. Though this practice improves boiler efficiency, re-injecting char tends to increase the percentage of small particulate. Figure 5-2[10] illustrates

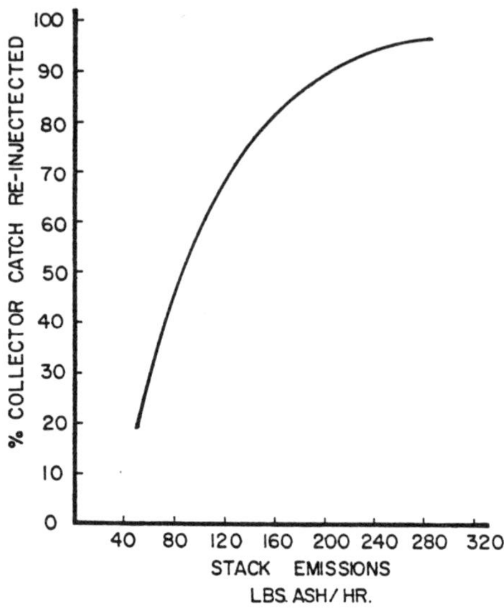

Figure 5-2. Effect of flyash re-injection on particulate emissions. (*from A. Barron, "Studies On The Collection Of Bark Char Throughout The Industry", Tappi Journal, vol. 53, August 1970.*)

the effects of re-injection on the particle loading. Note that 100% re-injection is not achievable since this would involve injecting already collected fly ash fines.

Boiler operation. Boiler operating methods and firing techniques have a direct effect on stack emissions. Unlike most natural gas or light fuel oil fired boilers, wood-fueled installations require full-time operator attention to achieve maximum efficiency. For example, the rate of flow of combustion air can be automated to the flow of natural gas or oil fuel, but this is not practical with wood systems. If the boiler fireman consistently supplies less air than is needed, unburned fuel will result and smoke will be produced. Excess combustion air will reduce combustion efficiency and increase the total mass of combustion products. This increased mass results in higher velocities in the combustion zone, which causes the entrainment of particulate matter and reduces the fuel residence time.

In addition to the quantity of combustion air, the method of its introduction can influence emissions. Wood combustion systems are often designed to introduce a percentage of the combustion air from beneath the grates and the remainder from above the grate line (under and over-fire air). The underfire air is necessary for combusting the fuel on the grates while the overfire air is used to combust the fuel volatile gases. While percentages of over-to-underfire air are a function of boiler design, a good fireman adjusts the air to meet the needs of particular fuels and operating conditions.

When a boiler is overloaded, a condition similar to that with high excess air develops. As the rated load is exceeded, air flow to the boiler is increased leading to higher velocities and more particle entrainment. Higher-than-normal emissions can be expected when a boiler is operated over its rating.

Upsets in the boiler operation can cause temporary excess emissions. Upsets can occur when the load changes rapidly or when the moisture content of the fuel varies unexpectedly. Though upsets with wood-fired boilers are difficult to eliminate, proper corrective action by the operator can reduce their effect.

Emission Factors

Emission factors for particulate emitted from wood-waste boilers without clean-up systems are presented in Table 5-1. The factors are presented as pound of particulate per ton of fuel burned. The factors for bark and bark mixtures are based on an as-fired moisture content of

TABLE 5-1. Particulate Emission Factors for Wood and Bark Combustion in Boilers.[10]*

FUEL	EMISSION (LB/TON OF FUEL—AS FIRED)
Bark:	
with re-injection	75
without re-injection	50
Wood/Bark Mixtures:	
with re-injection	45
without re-injection	30
Wood	5–15

*From AP-42, ''Compilation of Air Pollutant Emission Factors,'' U.S. Environmental Protection Agency.

50%. The wood fuel considered includes sawdust (5 to 50% moisture), shavings, ends, etc., and no bark. For well designed and operated boilers, use lower emission factors; and in the opposite case, use higher values. The emission factor for wood only is expressed on an as-fired moisture content basis assuming no fly ash re-injection.

The U.S. Environmental Protection Agency (EPA) compiles expected emission levels from combustion sources based on pounds of fly ash per ton of fuel burned. These factors can only be used for approximate calculations because they do not consider the effects of variables mentioned earlier.

Emission Characteristics

Several properties of particulate matter from wood combustion are important when considering control equipment. The particle size distribution is an important factor in control device operation. Smaller particles (less than 10 microns) are difficult to trap. Gas streams containing large percentages of sub-micron particles will require large energy inputs (that is, pressure drop) for cleaning. Fortunately, particulate from wood combustion is relatively large in size (Figure 5-3[10]).

Particle strength is also a property to be considered in collection. Carbon particles break very easily into smaller particles, making collection by mechanical means difficult. Therefore, the first-stage mechanical collectors used to capture unburned char are designed for low velocities. This helps limit shattering when impact occurs.

Particulate resistivity is an important property to consider if an electrostatic precipitator is to be used for collection. Particles must exhibit a relatively high resistivity (the ability to accept and hold a charge) to be collected effectively. Wood ash generally has a low resistivity—rendering it less effectively collected by electrostatic precipitators.

Particles that are adhesive may not be more difficult to collect, but are always more difficult to remove from the collection device. Adhesive particulate tends to clog up collection equipment resulting in decreased collection efficiency. Adhesiveness, however, is generally not a problem with particulate from wood ash.

CONTROL TECHNIQUES

Many devices are currently available for the control of emissions from combustion sources. The five major control devices in use today are

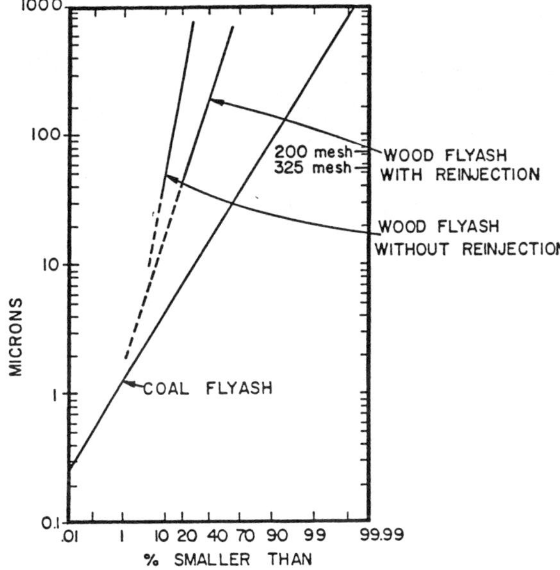

Figure 5-3. Size distribution of wood and coal fly ash. (*from "Comparison Of Fossil And Wood Fuels", U.S. Environment Protection Agency Report EPA-600/2-76-256, March 1976.*)

mechanical collectors, baghouses, wet scrubbers, dry scrubbers, and electrostatic precipitators. Table 5-2[10], "Relative Collection Efficiencies of Various Pollution Control Devices," shows the size range of particulate each will collect efficiently and economically.

TABLE 5-2. Relative Collection Efficiencies of Various Pollution Control Devices.[10]

| | COLLECTION EFFICIENCY % | | | | | |
| | PARTICLE SIZE RANGE, MICRONS | | | | | |
TYPE COLLECTOR	OVERALL	0–5	5–10	10–20	20–44	44
Simple Collector	65.3	12.0	33.0	57.0	82.0	91.0
Multiple Cyclone (12″ dia)	74.2	25.0	54.0	74.0	95.0	98.0
Multiple Cyclone (6″ dia)	93.8	63.0	95.0	98.0	99.5	100.0
Electrostatic Precipitator	97.0	72.0	94.5	97.0	99.5	100.0
Venturi Scrubber	99.5	99.0	99.5	100.0	100.0	100.0
Baghouse	99.7	99.5	100.0	100.0	100.0	100.0

Mechanical Collectors

The most common control device is a cyclone—or the centrifugal dust collector. Cyclones are used primarily to collect larger dust particles. In the period before the emphasis on emission control, their purpose was to clean the air stream of large particles in order to protect induced draft fans against erosion. Particles are removed through the action of a double vortex. The inlet gases spiral downward creating a centrifugal force which pushes the particulates toward the wall (Figure 5-4[17]). The particulate drops out as the gases change direction and spiral upward to the

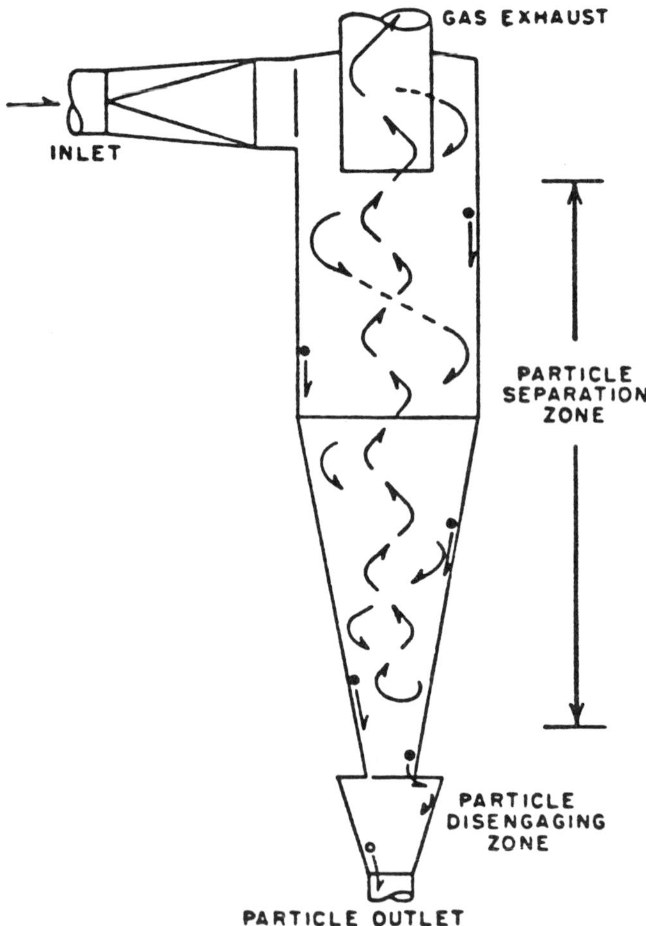

Figure 5-4. Cyclone collector.

exit. Cyclone collectors are best at removing large particles; their efficiency decreases as the mean particle size decreases. An important controlling parameter is the cyclone diameter. As the diameter increases, particles must travel farther to reach the wall; thus, efficiency decreases. This diameter limitation is overcome by using a multi-cyclone which is a series of small diameter cyclones arranged for parallel flow. Multiclones can handle the same air flow as large diameter singular cyclones but at a higher efficiency. The expected pressure drop through a cyclone or multiclone is 1 in. to 6 in. water column. If a mechanical cyclone collector is insufficient to meet emission standards, it is usually used as the first stage of collection to be followed by an appropriate secondary collector.

Baghouse

A baghouse is a container housing rows of cylindrical bags (Figure 5-5 [17]). Air can enter from the top or bottom. As air passes through the bags, particulate matter is trapped. The trapped particles are removed from the bags by reversing air flow, shaking, or impingement of a high velocity air jet. Baghouses are extremely efficient (above 99% even for sub-micron particles), have a low pressure drop, and require low fan horsepower.

Historically, baghouses have seen little application in wood-fired boiler applications because of the fire hazard created by spark carryover. Maintenance costs are high because the repeated flexing action to remove trapped particulate decreases bag life. Bags typically are replaced every 18 to 24 months. The bags are temperature limited to environments of less than 600°F. Among the bag materials in this temperature range are fiberglass with teflon coating, nomex, and felted teflon. A mechanical collector is used upstream from a baghouse to remove large glowing particles. Operating the baghouse at positive pressure to prevent air leakage into the baghouse limits the possibility of fires by keeping oxygen from igniting any glowing char.

Wet Scrubbers

Wet scrubbers are designed to develop an interface between a scrubbing liquid and the gas to be cleaned. The particles in the gas are trapped by liquid droplets; then the liquid is collected and removed. There are a variety of wet scrubber designs available—baffle and spray (Figure 5-

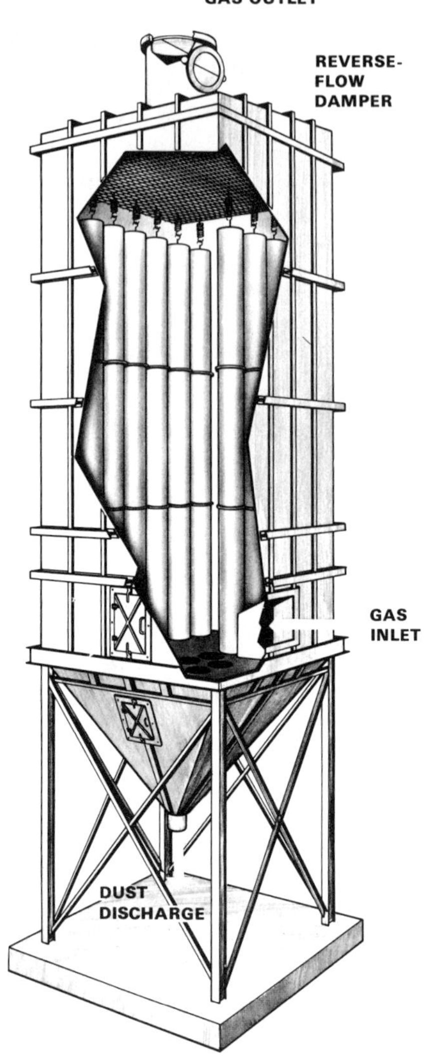

Figure 5-5. Fabric filter system. (*courtesy Zurn Industries Inc.*)

6[17]), venturi (Figure 5-7[17]), and impingement (Figure 5-8[17]) to name a few. The type of wet scrubber is chosen based on allowable emissions, particulate size distribution, and gas conditions going into the scrubber. In wood applications, usually baffle and spray-or-venturi types are used.

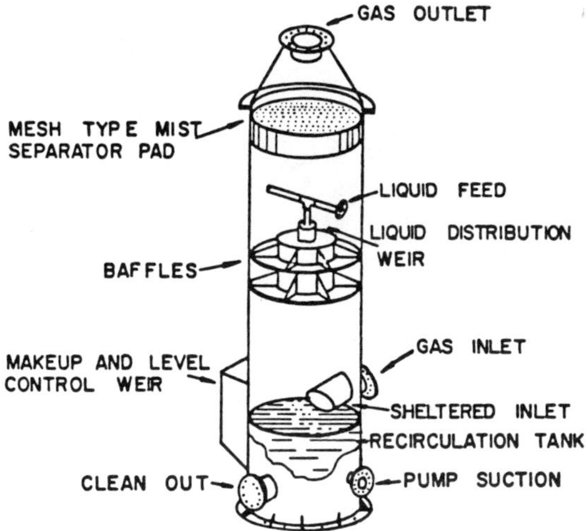

Figure 5-6. Baffle and spray. (*courtesy of Andersen 2000 Inc.*)

Wet scrubbers have the advantage of high efficiency (even on small particles), relatively low first and maintenance costs, and resistance to fire damage since glowing sparks are water-quenched. The obvious disadvantage of a wet scrubber is the problem of water supply and disposal. Wet scrubbers have a high demand for water (usually 5 to 10

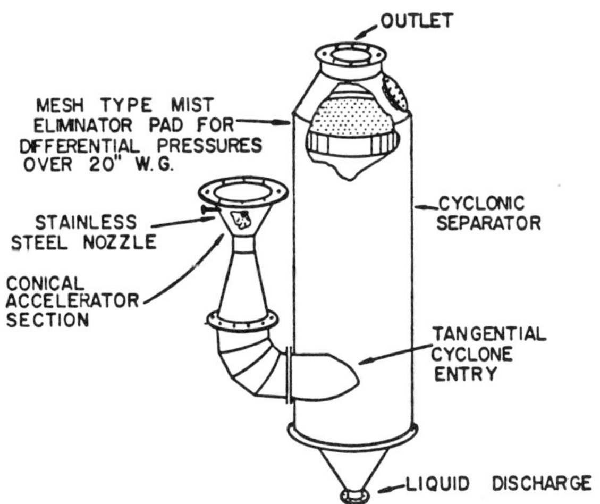

Figure 5-7. Venturi. (*courtesy of Andersen 2000 Inc.*)

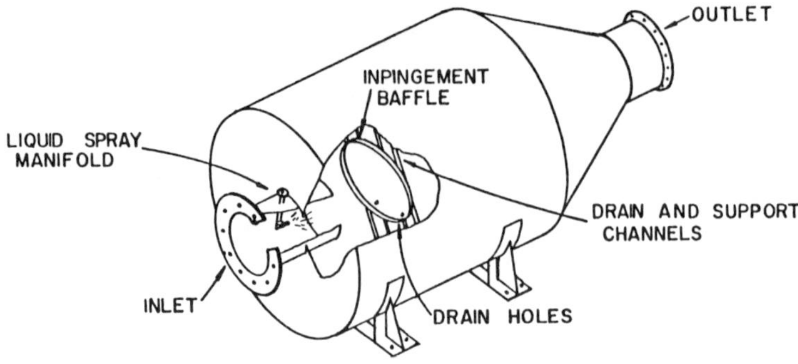

Figure 5-8. Impingement scrubber. (*courtesy of Andersen 2000 Inc.*)

gallons per ACFM). Once the liquid has collected the particles, it must be cleaned through the use of a clarifier or settling pond. Also, the energy requirements of a wet scrubber increase as mean particle size to be collected decreases. Most wood applications require a low-to-medium energy scrubber (6 to 15 in. water column).

Dry Scrubbers

Dry scrubbers offer an alternative to wet scrubbers in areas where no water supply is available or water pollution presents a problem. Dry scrubbers trap particulate matter in a moving bed of granular material. The trapped particulate material can be removed from the media, permitting its recycling. Efficiency and pressure loss are comparable to a wet scrubber, but no water is needed and the overall scrubber is smaller in size. Dry scrubbers have not seen wide application on wood-fired boilers. The few locations using dry scrubbers did so because a wet scrubber was not feasible. The major complaint against dry scrubbers on wood fired-boilers has been their plugging by condensed hydrocarbons, and tars.

Electrostatic Precipitators

Electrostatic precipitators (Figure 5-9[17]) operate by ionizing particles as they enter the devices. These charged particles are then attracted to oppositely charged collection plates. The plates are cleaned by periodical mechanical "rapping." Dislodged particles are collected in a hopper.

The collection efficiency strongly depends on the resistivity of the particles to be collected. If the electrical resistivity of the particles is extremely high or low, the efficiency of an electrostatic precipitator will be adversely affected. Because particulate emissions from wood boilers are generally low in resistivity, they are not collected effectively by electrostatic devices.

Electrostatic precipitators have seen wide application in installations

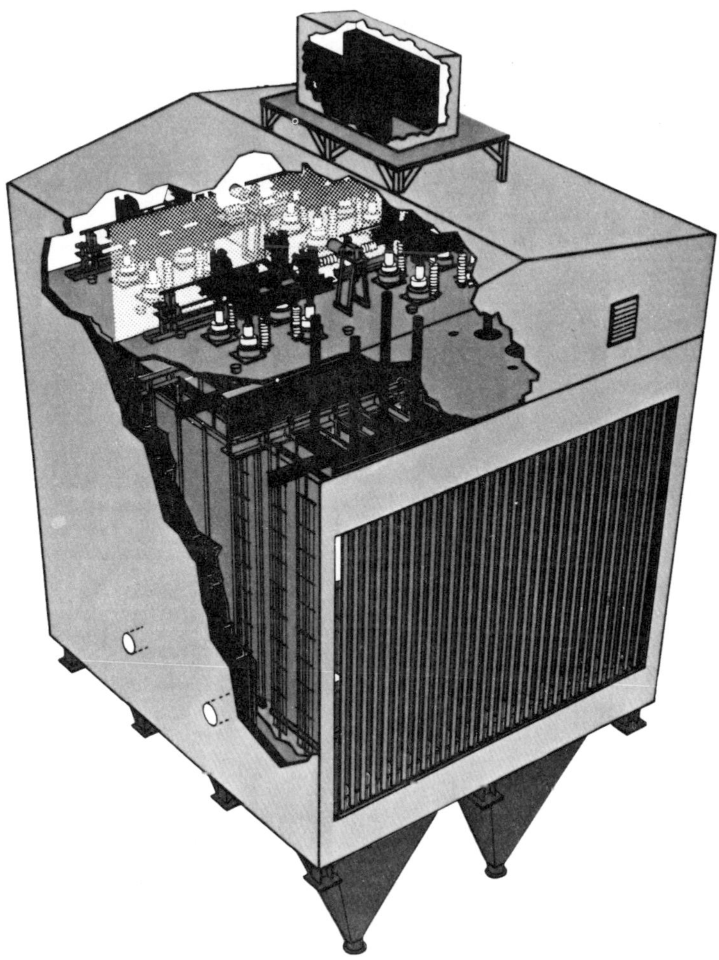

Figure 5-9. Large electrostatic precipitator. (*reprinted with the permission of Environmental Products Division, Dresser Industries, Inc.*)

where coal is burned and are used in applications where coal/wood mixtures are burned. Because of their uncertain effectiveness and high cost, they have seen almost no use on wood-fired boilers.

COSTS

Estimated costs per CFM of flue gas for various emission control systems are presented in Table 5-3. These are approximate installed costs for the aboveground structure. These costs are shown as a comparison between various control methods and should not be regarded as final, since site specific costs depend on the characteristics of each individual installation.

TABLE 5-3. Pollution Control Equipment Installation, and Operating Costs.[10]

	INSTALLED COST ($/CFM)	Δ P (IN H_2O)	YEARLY ENERGY COST FOR FAN OPERATION (DOLLARS)
Single Multi-Cyclone	2	2	4,000
Double Multi-Cyclone	4	4	8,000
Spray & Baffle Wet Scrubber	4	1–3	2,800
Venturi Wet Scrubber	6	10	10,000
Dry Scrubber	8	6	5,600
Baghouse	11	6	5,600
Electrostatic Precipitator	15	½	1,000

On smaller installations where the grain loading to the collector is relatively low and the emission regulations are not as stringent, two mechanical collectors in series, one after the other, may be sufficient to meet the standard. The first collector removes larger particles while the second collector is equipped with high energy vanes to remove the bulk of the finer material. The overall efficiency of such a system can be as high as 97 to 98%.

Systems with particulate loading too high to be collected with series mechanical collectors, but low enough not to require full air flow to be sent through a scrubber, can sometimes meet the standard with a selective collector-scrubber system. The full flow goes through a mechanical collector where the dirtiest portion (10 to 20% of the total) is separated out and fed to a wet scrubber. This reduces the size and cost of the wet scrubber to allow emission certification at the most economical cost.

Two different selective flow scrubber systems—shave-off and fractionating—are currently on the market. In the shave-off system, the multiclone outlet is designed with two outlet tubes instead of one. The inner tube allows the core of cleanest air to pass through. The outer tube shaves off the perimeter of the outlet gases where, due to centrifugal action, the majority of the particulate resides.

The fractionating system approaches the problem by removing the particulate before the airstream reaches the outlet tube. A manifold is attached to the cyclone below the tube sheet, allowing a controlled extraction of air from the large cyclone cavity. This extraction aids particle disengagement from the air when it changes direction at the exit, and the air bleed-off air tends to sweep the dust from the vessel bottom.

Most large scale operations require a mechanical collector followed by wet scrubber. Due to the tighter emission standards on large boilers, all the flow is sent through the scrubber. This combination should be able to meet even the most stringent emission standard.

The purpose of the selection process is to select the most economical control system which will still meet the regulations. In a turnkey operation where a design engineering firm handles the installation, compliance with emission standards is guaranteed by the vendor. The design firm and the boiler manufacturer meet, and by considering important criteria such as fuel type, boiler size, and emission control labels, arrive at an appropriate design.

Companies who buy boilers without the aid of a consulting enineer generally fall into two categories: (1) the small firm which buys a turnkey installation from a boiler contractor; and (2) the very large firm where all engineering is handled in-house.

In both cases, economics is the pre-eminent consideration in the selection of an emission control system. Buy only as much control as required. Know what fuel will be used, as emission control systems are fuel specific. Knowledge of fuel properties and proper boiler operating procedures will assure a boiler and emission control system design that meets the regulations as economically as possible.

Estimated cost information for several different size systems is presented in Table 5-4, and Figures 5-10[13] and 5-11[14]. This price includes the control equipment, connecting ducts, support steel, stacks, controls, settling pond (where applicable), and foundations. In general, emission control requirements are determined during the boiler design phase, and the cost of the necessary devices is included in the boiler package cost.

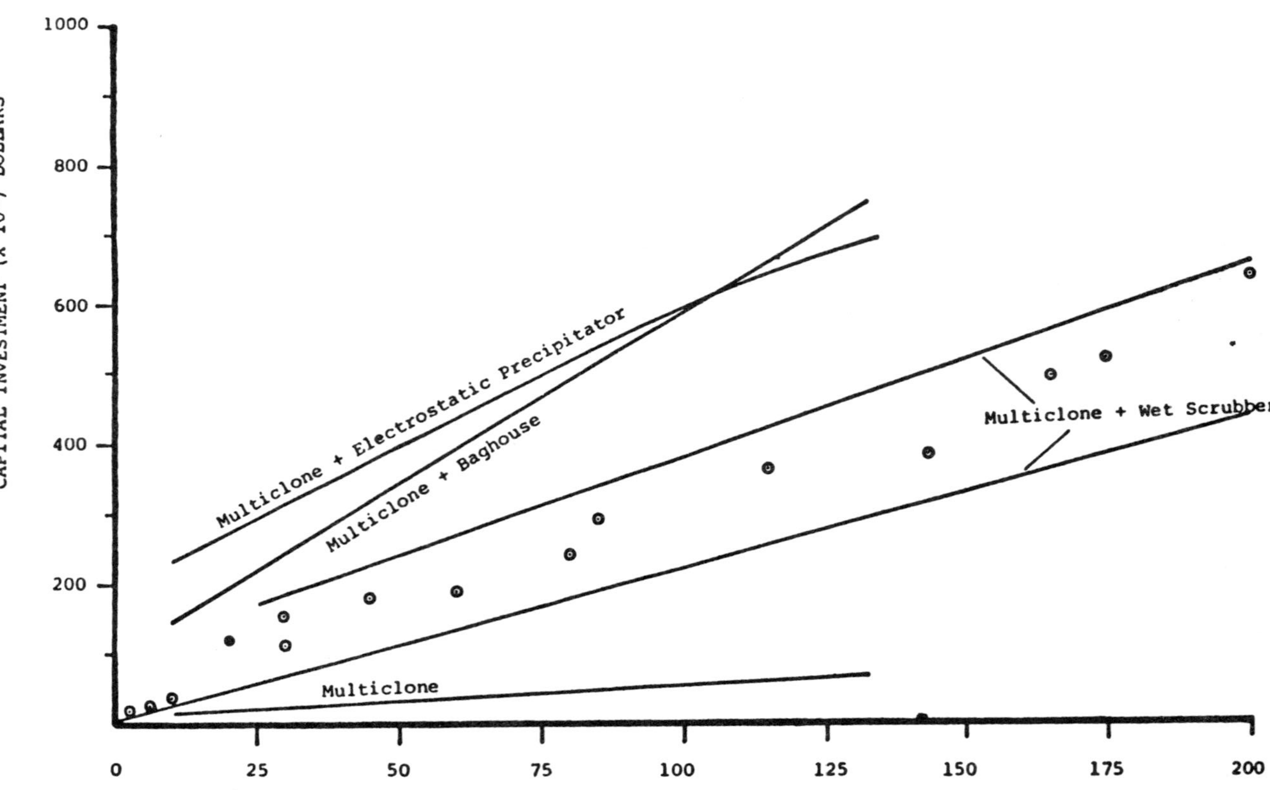

Figure 5-10. Capital investment for air pollution control.

TABLE 5-4. Pollution Control Systems (Complete): Installed Costs.[10]

BOILER SIZE (LB/HR)	SYSTEM	FAN SIZE (HP)	INSTALLED COST (DOLLARS)	EST. YEARLY OPER. COST (DOLLARS)
3,450 (100 bhp)	Multi-cyclone	5	16,500	670
17,250 (500 bhp)	Series Multi-cyclone	30	34,500	6,700
50,000 (1,450)	Multiclone/Wet Scrubber	150	150,000	70,000
100,000 (2,900 bhp)	Multiclone/Wet Scrubber	300	225,000	137,000

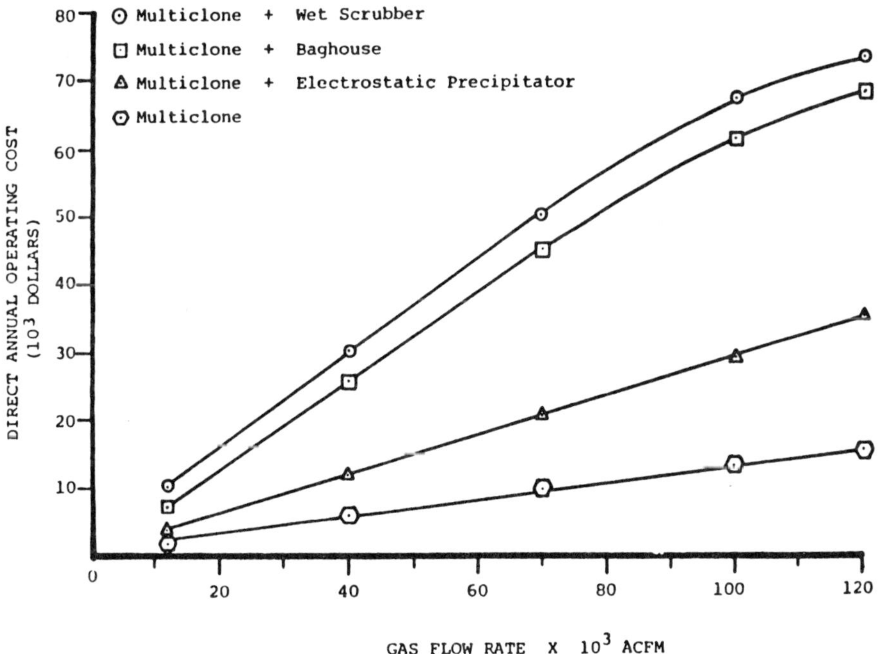

Figure 5-11. Operating costs of air pollution control systems.

The operating cost for multi-cyclones consists of the fan horsepower required to overcome the pressure loss. The costs were calculated assuming boilers operating at 100% excess air for 6,000 hours per year with electricity at 5¢/kwh. For a wet scrubber, the operating cost consists of fan horsepower, pump horsepower, and water used. The pressure loss assumed was seventeen inches of water. A 5% evaporative loss was used to arrive at the water cost. The pump horsepower was assumed to be 5% of the fan horsepower. Excess air, yearly operating time, and electricity costs were the same as used with multiclones.

6

Environment and Safety (Rules, Regulations and Safe Practice)

ENVIRONMENTAL IMPACT

Since most users first encounter the wood fuel when it is delivered to their plant site, the environmental problems presented stem from storage, handling, and use of the wood as a fuel.

During the past ten years, a number of studies have been made regarding the environmental impact on the growth and procurement of wood. These studies have considered harvesting operations both for timber as well as wood for energy. While the two operations are very similar, those that recover wood for energy usage recover a greater total portion of the tree biomass which in turn increases clearcutting and frequently reduces the rotational life cycle of the tree. The principal results of such practices are the elimination of animal habitat and forage, increased soil temperature, water runoff and erosion, soil compaction and aeration, and nutrient depletion.

With proper management, the harvesting of wood for energy will not impact environmental factors in any greater degree than harvesting of other timber products. Markets for energy wood could, in fact, help improve the environment by providing an incentive for conversion of unproductive forest stands and an opportunity for improved cuts in managed stands.

PERTINENT ENVIRONMENTAL REGULATIONS

The storage and handling of wood fuel is not normally environmentally regulated, with the exception of water runoff from outdoor storage piles discharged into running streams. But the combustion products—both airborne and solid waste—are of concern to the Environmental Protection Agency. There are four major documents that relate to these areas: (1) National Ambient Air Quality Standards as amended through 1971 (NAAQS); (2) Prevention of Significant Deterioration—Clean Air Act as amended through 1977 (PSD); (3) New Source Performance Standards (NSPS); (4) and Solid Waste Management Act as amended through 1973.

The National Ambient Air Quality Standards were developed in 1971 for six pollutants which reflect thresholds of atmospheric concentrations above which the pollutants are thought to have significant deleterious effects on human health and/or on plant and animal life and property. These pollutants are particulate matter, sulfur oxides, nitrogen oxides, carbon monoxide, photochemical oxidants, and hydrocarbons. The NAAQ Standards are expressed in the terms of primary and secondary standards. The primary standards are specified to protect public health, while more stringent secondary standards are set to protect against effects on soil, water, vegetation, materials, animals, weather, visibility, and personal comfort and well being. The primary standards are to be met in a reasonable time as determined by the Environmental Protection Agency. The NAAQ Standards are listed in Table 6-1.

A more recent regulatory development that has greater potential to directly affect wood-burning devices are the Prevention of Significant Deterioration standards. These standards state how much of an increase a single source can add to the ambient air quality of a particular region for particulate matter and sulfur dioxide. There are three classes of PSD standards. All areas of the country will be designated as either Class 1, Class 2, or Class 3 regions for application of PSD standards. Class 1 standards allow the smallest incremental decrease in ambient air quality. Congress has already designated specific regions as Class 1. These are areas of pristine air quality such as certain size national and international parks, and national wilderness areas. Areas designated for application of Class 2 standards are allowed a larger incremental decrease in ambient air quality, although not as large as allowed in the Class 3 areas. NAAQS will act as an overriding ceiling to any otherwise allowable increment (this includes all specified NAAQS pollutants and not just particulate matter and sulfur dioxide). This PSD statutory scheme will not be fully effective until the state and/or EPA undertake further rule-making activity. The PSD regulations state that all areas not classified

TABLE 6-1. Ambient Air Standards.[a]

	AVERAGING PERIOD[c]	AIR QUALITY STANDARDS[b]	
		PRIMARY[a]	SECONDARY[a]
SO_x (as SO_2)	AAM	80(0.03)	
	24	365(0.14), 1x	
	8		
	3		1,300(0.49), 1x
Particulate	AGM	75	60
	24	260, 1x	150, 1x
	8		
Carbon monoxide	8	10,000(9), 1x	10,000(9), 1x
	1	40,000(35), 1x	40,000(35), 1x
Oxidant (as O_3)	8		
	1	160(0.08), 1x	160(0.08), 1x
NO_x (as NO_2)	AAM	100(0.05)	100(0.05)
	24		
	8		
	1		
Nonmethane (HC)	8		
Hydrocarbons (HC)			
(as CH_4)	3	160(0.24), 1x	160(0.24), 1x

[a]Format for each entry is as follows: Standard $\mu g/m^3$ @ 760 mm Hg and 20° C (equivalent value, ppm). The maximum allowable exceedance rate, if any, follows. This refers to the maximum number of times per year that the standard may be exceeded. For example, 1x means that the standards may be exceeded only once per year.
[b]"National Primary and Secondary Ambient Air Quality Standards," *Federal Register 36,* #84, pp. 8186–8201.
[c]The averaging period is given in hours unless otherwise specified. AAM means Annual Arithmetic Mean Value and AGM means Actual Geometric Mean Value.
Source: Fennelly, P. F. et al, *Environmental Assessment Perspectives,* GCA Corp. for U.S. EPA, Washington, D.C., March 1976, p. 213.

as Class 1 will be designated Class 2. A reclassification process is involved in changing any area to Class 3. This reclassification involves among other items specific approval from the governor of the affected state after consultation with the state legislature and with local governments representing a majority of the residents in the area which is to be redesignated. Table 6-2 lists the PSD standards.

Major pollution emitting facilities are required to have a PSD permit before construction can begin. The permit acknowledges that all PSD emission requirements will be met by the facility; that the proper procedures have been undertaken—such as hearings, reviews, analysis of air quality impact, etc.; that the best available control technology for each pollutant is obtained; that proper air quality monitoring will be done; and so on. Twenty-six specific emitting facilities are defined in the act as requiring permits, if they emit 100 tons or more of any air pollutant. Also the PSD permit requirements include other facilities which

TABLE 6-2. Prevention of Significant Deterioration Standards.

| | MAXIMUM ALLOWABLE INCREASE* | | |
POLLUTANT	CLASS I	CLASS II	CLASS III
Particulate Matter:			
Annual geometric mean	5	19	37
24-hour maximum	10	37	75
Sulfur Dioxide:			
Annual arithmetic mean	25	20	40
24-hour maximum	5	91	182
3-hour maximum	25	512	700

* $\mu g/m^3$
Source: *Federal Register*, Vol. **42,** No. 212, 11/3/77, p. 57459.

are not specified but which have the potential to emit 250 tons or more per year of any pollutant. These permit requirements also apply to major modifications of existing facilities which cause emissions to increase by 100 tons per year or 250 tons per year depending on whether or not it is a specified or unspecified emissions source. Because of court rulings, potential emissions from a given source will be calculated after accounting for a reasonably anticipated effect of air pollution controls that are part of the facility design. Calculation of potential to emit will assume a round-the-clock operation at maximum rated capacity unless the applicant can show it is impossible for the source to operate in that manner. Thus, some new sources will no longer be required to obtain PSD permits if the emissions can be controlled to less than 100 tons per year if they are on the specified emission source list or 250 tons per year for the remaining sources.

Existing emission sources which are going to be modified with the result of new emission rates over the 100/250 tons per year limits will also require PSD permits. There are some exceptions to the above requirements. These exceptions deal with emission rates that are negligible and thus fall outside the strict requirements of the law. There are two tests for this. First, if mass emissions will fall below a certain rate, the source will not need a PSD permit but must notify EPA of the proposed construction. Second, if the mass emissions are above the specified rate but the ambient air quality impact of these emissions is below the significant impact guideline, the emissions do not need to be reviewed for air quality impact.

An analysis of the minimum size boiler that would fall under PSD is

TABLE 6-3. Minimum Size Wood-Energy Boiler Which Would Produce a Level of Emissions Requiring a PSD Analysis.

TYPE OF EMISSION	EMISSION FACTORS (LB/10^6 BTU)	SIZE OF BOILER REQUIRED TO PRODUCE:	
		100 TON/YR OF THE GIVEN POLLUTANT (LBS OF STEAM/HR)	250 TON/YR OF THE GIVEN POLLUTANT (LBS OF STEAM/HR)
PARTICULATES			
Bark:			
With fly ash reinjection	8.72	1,911	4,778
Without fly ash reinjection	5.81	2,869	7,172
Wood/Bark:			
With fly ash reinjection	5.23	3,186	7,967
Without fly ash reinjection	3.49	4,775	11,939
Wood	0.58–1.74	28,736–9,579	71,839–23,946
Sulfur Oxides	0.17	98,039	245,098
Carbon Monoxide	0.23–6.98	72,464–2,388	181,159–5,970
Hydrocarbons	0.23–8.14	72,464–2,048	181,159–5,119
Nitrogen Oxides	0.25	66,667	166,667

Assumptions: Boiler operates 350 days/year
1,000 Btu/lb steam
70% boiler combustion efficiency
50% moisture content
$8.6 \cdot 10^6$ Btu/ton of green wood

given in Table 6-3. Another set of pertinent regulations are the New Source Performance Standards (NSPS). These are basically standards that limit the amount of pollutants that can be emitted per million Btu of fuel burned. The present standards regulate emissions from coal-fired utility boilers that have an energy input greater than 250 million Btu/hr, produce electricity, and 25 megawatts or more of this electricity goes to the grid for commercial use. These regulations limit particulates, sulfur oxides (SO_x measured as SO_2), and nitrogen oxides (NO_x measured as NO_2) as follows:

Particulates. May not exceed 0.03 lbs/million Btu of coal burned.

Sulfur oxides. May not exceed 1.2 lbs/million Btu of high sulfur coal burned and a minimum of 90% of SO_x must be removed even though this places the emission well below the limit.

May not exceed 0.6 lbs/million Btu of low sulfur coal and a minimum of 70% of the SO_x must be removed even though this places the emission well below the limit.

Nitrogen oxides. May not exceed 0.6 lbs/million Btu of high sulfur coal burned and may not exceed 0.5 lbs/million Btu for any other type of coal burned.

Only when wood is burned in combination with coal do these NSPS regulations apply to wood combustion; specifically, if a utility boiler (a) produces 25 megawatts or more of power for commercial consumption, (b) burns a mixture of coal and wood of which the coal cannot exceed 25% of the total energy combustion in a 90-day period, and (c) has coal input exceeding 250 million Btu per hour, then the NSPS regulations for coal apply with the modifications that the SO_x limit is 1.2 lb/million Btu for all coal types and there is no percentage reduction requirement.

In a practical sense, there are few installations where NSPS regulations will apply. Standards for utility and industrial boilers that are 100% wood fired are forthcoming but not yet published. It is believed that these standards will likely focus on particulate emission. The magnitudes of particulate emissions are dependent upon the chemical and moisture content of the wood, the degree of fly ash re-injection employed in the boiler, and the boiler design and operating conditions. Control technology in the form of cyclones, wet and dry scrubbers, baghouses, and precipitators can collect up to 99.9% of the particulate matter emitted. Became of the low sulfur content of the wood, sulfur oxide emissions are minor. Nitrogen oxide emissions are also minor in a well operated facility. Therefore, if regulations are issued on SO_x and NO_x, they will most likely be a minor form and not of major concern.

The production of carbon monoxide and aromatic hydrocarbons rises sharply with poorly designed equipment and/or poor operating conditions. Therefore, attention is being given to the issue of including among the regulated pollutants carbon monoxide and hydrocarbons, particularly polycyclate aromatic hydrocarbons (PAH). PAH are of concern because given a large enough concentration some may be carcinogenic.

Solid Waste

Ash waste from wood combustion is considered less detrimental than solid residues of coal combustion, with the only toxic substance that has been identified in wood ash being lead—present in minor quantities. Ash from wood burning is, however, considered solid waste and as such is subject to the provision of a solid waste management act which is administered by the Environmental Protection Division of the Department of Natural Resources. This act provides the requirements for permits for solid waste handling and operation of disposal sites.

The solid waste management program is separated into municipal solid waste management and industrial and hazardous waste management programs. Industrial and hazardous waste management is concerned with the residual waste generated from industrial operations.

The compliance effort of the program determines waste streams and waste inventories, and monitors operation of facilities with emphasis being given to resource recovery, reclamation, recycling, and use of residuals to eliminate the residual waste. Industrial facilities that are designed to operate their own disposal sites make application to the division for permit. Upon receipt of the application, the environmental acceptability of the proposed site is determined. Then, upon completion of the assessment and review and approval of the design and operation plan, a permit is issued for the operation of the site. Requirements for disposal of residual ash waste are provided in the rules and regulations of the act. These rules and regulations utilize criteria equivalent to that contained in subtitle D of the Federal Resource Conservation and Recovery Act of 1976. Briefly, the criteria include the following:

1. Floodplains—facilities or practices in the floodplains shall not restrict the flow of the base flood (100 year floodplain), reduce temporary water storage capacity of the floodplain, or result in washout of solid waste, so as to pose a hazard to human life, wildlife, or land or water resources.
2. Endangered species—facilities or practices shall not cause or contribute to the taking of any endangered or threatened species of plants, fish, or wildlife nor result in the destruction or adverse modification of the critical habitat of endangered or threatened species.
3. Surface water—a facility or practice shall not cause: (a) a discharge of pollutants into waters in violation of the requirements of the National Pollutant Discharge Elimination System (NPDES) permit; (b) a discharge of dredged or film material in the water in violation of Section 404 of the Clean Water Act; (c) non-point

source pollution of waters violating Section 208 of the Clean Water Act.

4. Ground water—a facility or practice shall not contaminate an underground drinking water source beyond the site boundary.

Runoff from open storage areas may occasionally be deemed a source of contamination to running streams; and to a lesser extent, ground water. The absence of a sloped concrete pad beneath the pile allows a depression to form under the pile. In this area, acidic water collects (pH of approximately 5) containing lignins and other water soluble components of the stored wood fuel. After a heavy storm, the collected water (bearing high contaminant concentrations) is flushed from beneath the pile. This solution, because of its high acidity, may be damaging both to ground water aquifers and aquatic environments. Provision of a sloped concrete pad beneath open storage areas will minimize problems of contaminated runoff.

SAFETY

The two major divisions of safety are safety of personnel that are involved with the fuel supply and the safety of the fuel supply itself. Personnel safety will come mainly under the jurisdiction of OSHA (Occupational Safety and Health Administration) which is a regulatory agency with fine assessing capability. Fuel supply will be under the jurisdiction of codes such as those promulgated by the National Fire Protection Association and the National Electrical Codes, which of themselves are not regulatory agencies.

Personnel Safety

OSHA was created by the Occupational Safety and Health Act (Public Law 91-596). This law is considered landmark legislation because it is the first national safety law in the history of the nation. It establishes standards which require each employer to provide his workers with workplaces free of recognized hazards that could cause death or serious injury. It also requires the employee to comply with all safety and health standards that apply to his job. Most of the rules and regulations that apply to personnel involved with a wood fuel supply are contained in "General Industry Safety and Health Regulations" Part 1910.

The applicable rules and regulations will depend upon each specific site, how extensively the wood fuel is processed, and the extent of the handling necessary to convey the fuel to the furnace. However, there are eight basic categories of injuries. These are: (1) electric shock or electrocution; (2) burns; (3) overexertion; (4) poisoning by inhalation, ingestion, absorption, etc.; (5) being caught in; (6) being struck by; (7) striking against; (8) and falls, both from the same level and from a different elevation. The objective of applicable rules and regulations is to provide equipment and processes which eliminate the potential for any of these eight types of accidents.

As a first step, all mechanical power transmission devices must be guarded—gears, chains, sprockets, belt drives, etc., should be enclosed. As a test for adequacy: wherever the power transmission drive may be located, if you can reach over, under, around, or through the enclosure and contact moving parts, then the guarding device is not adequate. Make sure that the machine is off before you try this!

The same criteria can be applied in determining whether the equipment's point of operation hazard is effectively guarded. If a hand or finger through any route can reach an area where work is performed on a material, then the potential for an accident exists and the machine is not adequately guarded. OSHA 2057 discusses the principles and techniques of mechanical guarding.

All equipment both stationary and mobile must meet certain standards. For example, the cab area of a front end loader must be protected by mesh screens that prevent entry of obstacles. Also, crane and loader stability must be in accordance with the American National Standards Institute's (ANSI) safety code for cranes, derricks, and hoists. Chippers should not be opened until a drum or disk is at a complete stop and this should be accomplished by interlocking. Machines for hogging wood should be so designed that at no position on the rim of the chute should the distance of the blades be less than 40 in. A safety belt and lifeline must be worn by any workman at or near the spout unless the spout is guarded.

Personal protective equipment must be provided for employees. These include iron face protection, safety helmets, safety shoes, gloves, and dust masks and hearing protection if the conditions warrant. Also operator instruction programs for the safe operating limits of the equipment are required.

Depending upon the size of the operation, storage and conveying systems can be either elaborate or simple. In all cases, though, there will be regulations governing work in and around storage bin areas, conveying equipment, and guarding of their ladders, walkways, and moving and rotating equipment. In addition to OSHA requirements, the American National Standards Institute directly addresses material handling systems in its "Standards for Conveyors and Related Equipment." This document deals with guarding moving parts, types of guards, control functions, platforms, operation, and maintenance procedures.

An aspect of safety that is addressed both by OSHA and insurance carriers is that of operator education and maintenance personnel instruction. Once equipment has been engineered to comply with safety standards and has been set up correctly, nearly every accident that occurs thereafter can be traced to the unpredictable element of the system—the people. It requires constant continuous management to ensure that the operators adhere to operating procedures and that maintenance personnel do not take shortcuts in their work.

In summarizing personnel safety, it is necessary to analyze the equipment, the system, and the plants for the potential of any of the eight different accident types occurring. Where the potential is found, the hazard must be eliminated with designs that use OSHA standards as guidelines and sources of technical information. And, as safety can't be legislated, safety education programs must be provided on a continuing basis.

Fuel Supply Safety

The major threat to wood fuel is fire. Fires can originate from both external sources and from internal combustion. Fires from external sources (such as smoking, electrical shorts, etc.) can occur in any size fuel wood. Internally initiated fires can occur in bark piles (see chapter 3) and piles of very small chips and/or sawdust (fines) since the internally generated heat is a function of the surface area/mass ratio and packing density. Whenever highly resinous dry fines are mixed with wet chips the potential for spontaneous ignition occurs. There are instances of chip piles recording temperatures near the boiling point of water and subsequently experiencing spontaneous combustion.

Recommended safe practices for storage and handling of wood chips and logs have been published in the National Fire Codes of the National Fire Protection Association. Experience has shown that the principal factors that allow lumberyard fires to reach serious proportions are large undivided piles, congested storage conditions, delayed fire protection, inadequate fire protection, and ineffective firefighting tactics.

A positive fire prevention program is required and should include selection, design, and arrangement of storage yard area and handling equipment based upon sound fire prevention and protection principles, facilities for early fire detection and transmission of alarm, fire extinguishers, fire lanes to separate large piles and provide access for effective firefighting operations, separation of yard storage from mill operations, regular yard inspection by trained personnel including an effective fire protection maintenance program, and a fire hydrant system connected to an ample water supply. Also pile height should be limited as heights in excess of 20 ft. seriously restrict effective extinguishing operations. Of course, smoking, cutting, welding, open fires, as well as unauthorized persons, should be prohibited and all electrical equipment and installations should conform to National Electrical Code or National Electrical Safety Code.

Storage precautions that may be taken to guard against internally generated fires in chipped or hogged wood piles are: keep all refuse and old chips out of the pile base; keep whole tree chips free from bark chips; use whole tree chips first; choose a clean storage site; keep buildup and reclaiming of a wood pile to a maximum turnover time of 1 year; limit pile size; install thermocouples during pile buildup or provide other means for measuring temperatures within the pile; avoid concentration of fines; wet the pile in dry weather; minimize vehicle compaction in the chip pile; and provide covered storage areas with automatic sprinkling protection.

Another aspect of wood fuel safety relates to combustion. Conventional fuels such as coal, natural gas, and oil have a relatively constant heating value. Consequently, it is practical to establish a desired fuel feed rate and to supply the proper amount of air for stoichiometric combustion.

Wood presents a more difficult problem in that its Btu content for a given volume or weight of fuel can vary widely. The species of the wood and percent of the wood bark in the chips will affect the heat content; however, the major factor is the amount of moisture the wood contains.

Establishing the correct air to wood fuel ratio is performed manually by the boiler operator by observing the flame. Wood fuel feeding equivalent is based on volume. For example, an auger feed unit will deliver a certain volume of material at a given rotational speed of the auger. The boiler operator soon learns the setting of combustion air flow for given auger speed. This technique works with reasonable satisfaction as long as the wood fuel is relatively uniform in its characteristics. A safety problem can be created if the fuel characteristics change suddenly.

If a system is running on green wood and dry fuel is introduced, the volume of material will remain constant, but the heat content can nearly double. To combust this dry material correctly, the volume of fuel should be reduced by ½ while keeping the combustion and air volume the same. In practice, a boiler operator may not realize he has a slug of dry material. Since there is not sufficient air to completely burn the wood the result is that the combustion chamber, the heat transfer section, and even the breeching chimney or boiler may be filled with a combustible gas. This condition is not unlike the analogy of having a leaking gas valve which fills the boiler with natural gas. A boiler filled with wood gas presents a potentially explosive situation.

There is presently no remedy for this condition except operator training and vigilance. Certainly, new boilers should be observed carefully during the startup phases and any time a new supply of fuel is received. While major explosions are quite rare, minor ones do occur and result in personal injury and equipment damage.

Other items of safety concern are fires in emission control equipment due to the accumulation of unburned carbon; spark carryover where burning wood particles are carried through the boiler and out the furnace stack; ashes which contain hot, glowing particles which are dumped on combustible materials; and housekeeping fires that occur from wood dust accumulation.

Section II

7
Feasibility Study Methodology

An engineer or manager performing a feasibility study for wood energy is often aware of the many aspects to be considered. The same person may be equally uncertain as to where to begin the study and in what order each aspect should be investigated. The purpose of this chapter is to present a methodology which organizes the data collection process in a flow that has been successfully used in performing many wood energy feasibility studies.

The step-by-step guide presented will be useful for those totally unfamiliar with wood energy as well as for those persons who have already been involved in studies. The flow (of the data collection steps) is set up so as to improve the efficiency of the study by limiting the collection of data until certain go/no-go indications have been passed. Effort expended in analyzing copious amounts of data is thereby prevented.

This chapter covers only the broad steps required to complete a study and their respective order of processing. Details relating to individual systems and wood fuel characteristics can be found in other chapters of this handbook.

The flow chart shown in Figure 7-1 depicts the order of the steps to be completed. The starting point must be a detailed assessment of the energy requirements. This assessment will be used to determine system characteristics, fuel type, fuel supply requirements, and so on. Therefore, it must be detailed in both short and long term aspects.

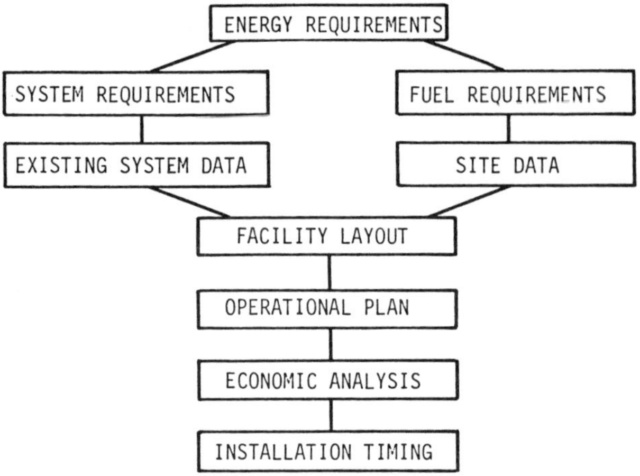

Figure 7-1. Order of investigation for feasibility studies.

ENERGY REQUIREMENTS

Energy requirements can be divided into three general categories:

- Load Cycle
- Hours of Operation
- Annual Consumption

Each category will be used for different aspects of the study. Abnormalities in energy demand should be equally detailed.

Peculiar to each individual operation, the load cycle has no defined time base. It can be as short as a few hours or as long as several months.

Consider a textile drying operation, as it might be represented by Figure 7-2. For simplicity, energy demand is shown as a percent of available capacity. In detailing your particular load cycle, however, a more quantitative form (such as pounds per hour of steam) should be used. For this example, a fairly constant base load is experienced with a periodic peak load—say from the opening and closing of a steam valve for a dryer.

Figure 7-3 depicts a demand cycle having a long time base. A common situation for industrial processes having some peak modulation of a non periodic form, the overall average load remains relatively steady,

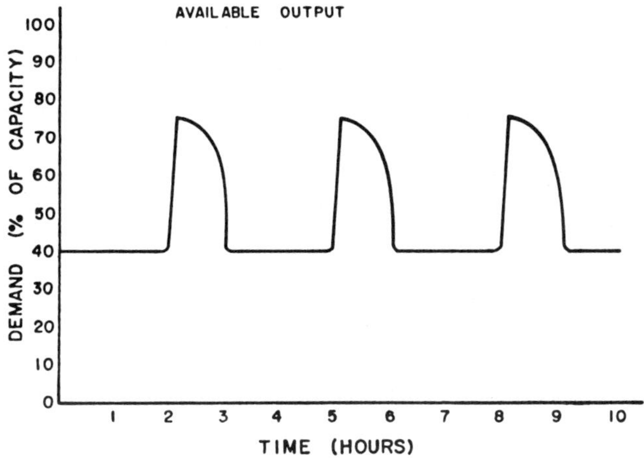

Figure 7-2. Industrial process steam with peak demands.

greatly exceeding the peak fluctuations. In this case, the average load would be used to represent the load cycle.

Figure 7-4 is an iteration, typical for space heating over a given season. The load cycle occurs on a daily basis, but from season to season it shifts over a wide range of demand.

Annual consumption can be determined directly by integrating the load curve (if the load cycle is uninterrupted). The next data to be tab-

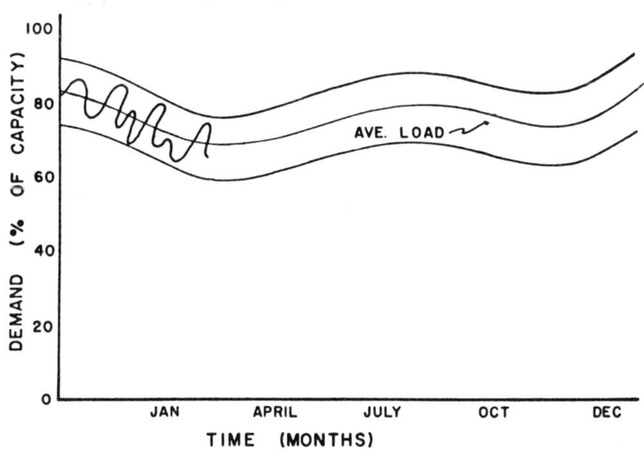

Figure 7-3. Industrial process steam with steady demand.

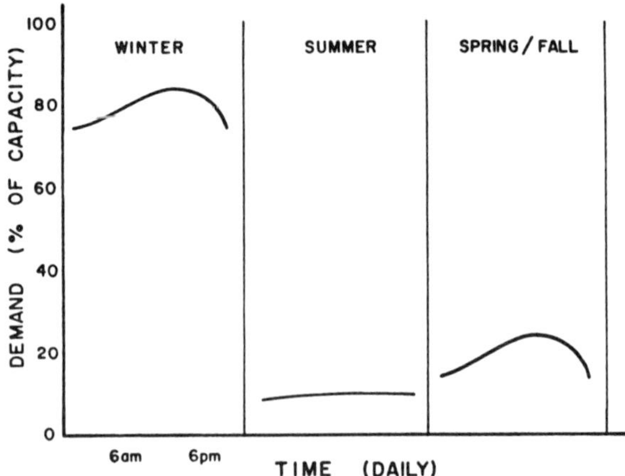

Figure 7-4. Space heating/hot water, seasonal demand.

ulated is the hours of operation. It must be defined not only for calculating the fuel requirements, but for the delivery schedules, materials handling, fuel storage, and so forth.

Having detailed the load curve, hours of operation, and any other abnormalities peculiar to the operation, the total annual energy consumption is calculated. After completion of the energy requirement assessment, it will often be necessary to proceed to the next step before a conclusive decision can be made.

The energy assessment gives insight into several areas. Any plant contemplating an alternate energy facility must operate a large percentage of the available time. This is true because savings to pay for the new system are generated by the use of less expensive alternate fuel.

Using fuel by operating more hours produces more savings. A large expenditure for an alternate energy system which sits idle is a waste of capital. Wood energy systems are most practical, then, in applications operating 7000 hours or more per year and least practical in applications operating 3000 hours or less.

For the same reasons, the size of the system is important. Large systems can produce more savings by using greater amounts of low cost alternate fuel. There will be a lower limit on the size of wood fuel systems below which economical justification cannot be made.

The load cycle could also influence the feasibility of the system. Wood systems, due to the thermal merits of the furnace refractory, tend not to adjust to rapidly changing load patterns. Where swing loads are anticipated, the best arrangement is to use a small base loaded wood boiler with a gas/oil boiler supporting the load swings.

SYSTEM AND FUEL REQUIREMENTS

Energy requirements can be translated into system and fuel requirements. If the investigator has little or no experience in wood fuel characteristics, assistance will be required. Information on wood handling, storage, supply, and combustion is presented in other chapters of this book. There are some "rules of thumb" calculations and charts to translate the energy requirements into fuel requirements and systems considerations (see chapter 3). Generally, a boiler efficiency of 65% is used. The fuel requirement for a 500 hp boiler is (assuming green wood fuel at 4500 Btu/lb × 2000 lb/ton = 9,000,000 Btu/ton):

500 hp × 34.5 lb/hr/hp × 1000 Btu/lb × 1/.65 × 1 ton/9,000,000 Btu
= 2.95 tons/hr of green (50% moisture content wet basis) fuel

Thus, a 500 hp boiler would consume approximately 2.95 tons/hr operating at full load. This assumes that 1000 Btu/lb are required for boiling. This approximation circumvents laborious, theoretical analyses and is practical for feasibility determination.

On the system side, a determination is needed of the boiler or furnace size necessary to meet the peak and annual energy requirements. The type of unit selected must be able to respond to the load cycle. The fuel feeding requirements will also evolve from this step. The energy requirement details can be translated into annual wood consumption, daily and weekly requirements, and storage or standby requirements.

Completion of this step will yield some solid go/no-go indicators such as:

- Suitability of a wood system to the plant load swings
- Potential savings produced by the energy demand
- Compatability of wood fuel delivery with hours and type of operation
- Availability of the wood fuel supply

EXISTING SYSTEM AND SITE DATA

If the system and fuel requirements seem plausible, the next step is to develop the available options by examining the existing system and site in detail.

On the system side, the possibilities of boiler conversion must be explored. Moreover, for conversion to be a real option, the system must be able, or have been able to, burn solid fuel. Solid fuel furnaces are equipped with the large furnace volume and floor grates necessary to support solids combustion. Package gas/oil boilers do not have the combustion volume nor the grates necessary to support wood firing. If the existing boiler is amenable to conversion, remember that boilers designed for coal firing will experience some degree of derating when converted to wood fuel (due to the larger volume of off-gases generated by combusting *moist* wood fuel). Generally, the boiler percentage derating will range from 25 to 30%. If the ability to convert is lacking, an entirely new piece of equipment must be considered.

Confirmation of the fuel supply requires contact with the local forest industry. Most plant personnel should be familiar with sawmill operators, loggers, and/or chippers in their area that could serve as fuel suppliers. Additional data may be gained by contacting the State Forestry Commission (see Appendix 2).

For the site, items to be investigated include confirmation of the fuel supply, transportation alternatives, receiving area configuration, storage alternatives, and overall available space. When contacting potential suppliers, it may be possible to learn if and how wood is being transported at present. Some areas have truckers with dump trucks or live bottom vans in operation. In most areas, wood is handled in conventional ''rag top'' trailers. The available method of fuel transportation will add insight into the type of receiving area necessary. If self-unloading alternatives such as live bottom vans are available, the fuel receiving area will be greatly simplified. Otherwise, a fuel receiving area must be designed. Its specification will depend on the amount of wood to be received. Small systems will be most economically served with a ''bobcat'' or equivalent front end loader, while large systems will best be served by a truck dump. Storage alternatives, discussed earlier in this handbook, consist of selected amounts of open and covered supply. The wood storage must be consistent with the plant characteristics and space availability.

The question of space is of prime importance when considering a wood energy system. Not only must there be sufficient space for the new boiler house, but storage space and vehicle accessibility must be provided. In locations where space is limited, a silo provides the best alternative for covered storage. Open storage can then be placed at some remote location.

Any number of go/no-go indicators may surface from these investigations and there is no defined order for the data collection during this step. Most investigators tend to examine fuel supply and storage areas first, probably out of greater concern for fuel availability. From a financial viewpoint, the wood supply and cost may provide an excellent benchmark in fuel cost savings, projecting a maximum capital outlay based on company R.O.I. and payback policies. This in itself is a good go/no-go indicator. Yet the examination of storage space is also important since it constitutes the largest area required by the system.

FACILITY LAYOUT

If indications are good, it is now time to make selections from all of the options and prepare the proposed facilities layout. This conceptual design should have concrete reasons behind each option selected such that if a problem is encountered later in the study a possible second alternative can be re-evaluated. A complete breakdown of hardware and implementation requirements will be needed for the following steps.

The facilities layout should cover every aspect of the system from wood supply through pollution control. With confirmation of a fuel supply, the delivery and receiving modes can be selected. This should include transportation to the site as well as unloading equipment, transport to storage, and so on.

If fuel processing is called for, the hogging and screening equipment and associated handling should be detailed. All storage areas should be located and sized to meet peak and contingency requirements, and materials handling in and out of storage should be specified.

The combustion system is based on either a conversion of existing equipment or a new installation. Ash removal should not be overlooked, and plans and equipment should be selected.

Pollution controls and alternate back-up fuels, if included, should also be incorporated into the proposed plan.

OPERATIONAL REQUIREMENTS

The operational plan will go hand-in-hand with the facilities layout and will include personnel additions, maintenance considerations, spare parts, and back-up requirements. Unlike gas/oil systems, a wood system will require a full time operator.

ECONOMIC ANALYSIS

Next, it is time to develop cost estimates of the proposed facility layout. Having assimilated all the costs, a full lifecycle analysis can be made. This can be as simple or detailed as company policy dictates. The payback and R.O.I. generated will most likely be the most significant go/no-go indicator, along with the total capital required for the proposal.

INSTALLATION PLANNING

The final step which may or may not be necessary for the particular study (but which makes an excellent complement) is an installation timing investigation. Part of the facilities layout will have addressed various pollution control options, and approval of permits can require healthy lead time. The timing of permit applications, shutdown scheduling, construction time, and equipment delivery lead times will be important factors if the system is implemented.

8
Wood Fuel Supply and Purchasing

DETERMINING WOOD FUEL SUPPLY

The cost of wood fuels varies from inexpensive mill residue (chips, sawdust, bark, and shavings) to dry wood pellets. Price is directly related to the type of wood (softwood, hardwood), moisture content, hauling distance, local supply and demand situation, and quantity purchased.

This section explains how a survey was taken to determine the mill residue supply that exists in Georgia; however, *the methodology is applicable to any locale.* There are references that give a methodology for determining the mill residue produced given a quantity of production, but this methodology is largely dependent on doing a study on the mill in question[1]. As more mill residue is utilized and sold, mill operators are becoming more aware of how much residue their mill produces and how much it is worth. The data base in this mill residue area is becoming more substantial.

At this time, mill residue is the cheapest fuel wood available. The variability of the material (sawdust, chips, boards, blocks, and slabs) and high moisture content (50%) may make handling, storage, and burning of this fuel costly. But its low price (8 to \$16 per ton delivered) can make it an attractive fuel. One particular study says only 8% is unused[2]. The majority of it is burned as boiler fuel or is used in making fiber products.

As a first step in establishing a data base for wood fuel supply, it is necessary to evaluate the many sources currently available. Generally, there is much information already assembled that can be used to good advantage.

The first source to check is the State Forestry Commission and the USDA Forestry Division. Part of their service is to maintain wood data. Most of the data that they publish concerns the amount of growing timber and removals on a county by county basis. This can give a good estimate with regard to the amount of mill residue or whole tree chips produced within a certain area. This does *not* necessarily mean that the number indicated is what is available for sale as fuel. It has been found that many of the operations are beginning to use more and more of the waste product for the purposes of steam production, heating, or drying their own products. Still, many are giving much of their residue away or carting it to the dump. Of course, producer utilization depends on the market conditions existing in the area.

A good place to look for suppliers of mill residue is in the State's manufacturing directory. This directory is probably located in the local library, or is obtainable from the State Department of Industry & Trade. It is much better to look at a current edition because of changes in ownership and fluctuations of the forest product industry's business with the economy. The companies in the directory are listed several ways, but the best for this purpose is the listing by Standard Industrial Classification (SIC) code. These codes are listed and defined in the manufacturer's directory. The SIC codes related to wood fuel producers are:

SIC CODE NUMBER	CLASSIFICATION
2421	Sawmills and planing mills—general
2426	Hardwood dimension and flooring mills
2431	Millwork
2448	Wood pallets and skids

There are other places—such as pulp and paper plants—that produce residue, but generally they use all that is produced and buy more. These four codes are the ones that we have addressed and from which a good reply was received.

Supply planning should be limited to areas that may be capable of supplying the quantity necessary for an industrial or commercial operation. There are three basic types of wood fuel for industrial fuel supply survey purposes: whole tree chips, wet or dry mill residue, and wood pellets.

The whole green tree chips are produced by running the whole tree through a large, expensive chipper. Few of these are operating currently in Georgia because of the current market situation. They do, however, produce many tons of chips per year and can be considered a good source of fuel. (The ones producing whole tree chips and interested in being a source of wood fuel are noted in Appendix 3. These are determined through our own surveys.)

Wood mill residue can be found at many sawmills, furniture manufacturers, pallet manufacturers, plywood and other forest product industries. Mill residue is a difficult entity to quantify—especially if the mill operator is not currently selling waste. He knows he has a huge pile of it every day, or it's stacked up out back, but he usually has only a vague idea of how much he has (or is producing) unless he's already selling it on a contract basis.

Wood pellets are not produced in many areas at this writing, but they may be produced in your area. It appears to be a fuel that is currently at a standstill because the cost of petroleum and competing solid fuels is not quite high enough yet to justify pellet cost. It is possible that price parity between densified wood fuel and petroleum fuels will be achieved over the coming years.

PURCHASING AND CONTRACTING FOR WOOD FUELS

After conversing with mill residue and whole green tree chips suppliers, it was found that contracts and specifications ran the full spectrum. Most everyone polled had some sort of verbal agreement that had been in force for some time. Others had written contracts which were 3 to 4 pages all the way up to 35 pages.

Most of the written contracts in the mill residue area were called waste purchase agreements. They ran from 1 to 5 years maximum. As a contract, they were loosely written. For example, they were reviewed semi-annually and could be changed within 30 to 60 days if both parties agreed; otherwise, the contract was void. These contracts generally contained f.o.b. price at the mill and a freight rate allowance. Both of these were subject to change at any time with the freight rate indexed to the prevailing freight rate.

The whole tree chippers worked both from a verbal (purchase order agreement) and a written contract. The verbal contract with the big mills was most prevalent. The whole tree chippers were in the 100-ton/day neighborhood and could more easily guarantee the supply to meet the demand without the need for the written contract.

The wood brokers which were contacted also used verbal and written agreements. One wood broker said that there was no standard contract. He said every one was different. The one common factor was that they were long—about 30 or more typewritten pages.

Another wood broker was not using a written contract at present, but was considering going to one. He also felt that each contract would have parts to it that would make it unique.

Overall, it appeared as though two important underlying factors were having a large impact on the use of written contracts. The two factors, supply and cost, are a result of the volatility of the wood market.

A sample proposal and an actual written contract are presented in the proceedings of the FPRS conference on "Hardware for Energy Generation in the Forest Products Industry" held in Seattle, WA, January 30–February 1, 1979. It is entitled "Specifying and Contracting for Wood Fuels," by Billy E. Wilson, B.G. Wilson Lumber Company, Hot Springs, AK.

Both the buyer and the seller of wood fuel are afraid of long term, fixed price, or projected price and supply contracts. They do not want to become involved in any handling contracts because of the uncertainties of supply and cost. It is because of this that the lumber and wood fuel business is one of the last to still operate by a "gentlemen's agreement."

9

Economic Analysis for Wood Combustion Systems

The economics of wood combustion systems are addressed by analysis of those factors common to any financial investment. The analysis focuses on the following topics:

- Capital Investment
- Operating Costs
- Taxes and Insurance
- Fuel Costs
- Annual Income/Savings
- Payback and Return on Investment

Budget capital costs for a proposed system are obtained by surveying vendors of wood fuel equipment and from estimations derived from construction handbooks. Major items are budgeted with installation included. All items include a full complement of accessory items and are "turnkey" estimates ready for operation.

Operating costs are the most difficult to estimate; however, they have minimal impact on the analysis. Wood fuel systems generally do experience higher operating costs because of their lesser degree of automation and increased personnel requirements. One successful approach to estimating operating costs is derived by surveying users of various systems. From the compiled cost data, a generalized cost is then estimated on the basis of Btu output.

Property tax and insurance costs vary depending on location, rate classifications, carriers, and so on. Specific detailed information should be obtained in this area when final design is initiated. Again, the survey approach can be used for this estimation.

Fuel costs are separated from the operating costs, as they have the greatest impact on the analysis.

The cost information assembled, the analysis can be performed on the basis of annual cash flows, taxes, etc. If conventional fuel systems and their respective costs are being displaced, savings (rather than income) are generated.

TAX CONSIDERATIONS

Tax consequences have a significant impact in business investment. An effort needs to be made to approximate these effects. Investment Tax Credit and Depreciation Schedules should be calculated. Federal and State taxes should be applied to the gross income after deducting depreciation. Investment tax credits can be applied to the ordinary tax liability. It can be assumed that losses created by tax credits or depreciation will provide tax relief for other corporation income and therefore can be treated as income (as opposed to carrying the loss forward to later years). It also should be assumed that the maximum tax rate applies to the project even if income is below the maximum tax levels, as other corporation income would be involved.

Depreciation

The depreciation schedule should be based upon current IRS guidelines for business investments and expenses. Under the present rules, tangible property used for the production of income may be depreciated, at a rate of 200% using the declining balance method. For simplicity, the entire investment capital may be considered as 20-year-life tangible property. In a more detailed analysis, shorter life equipment such as front end loaders, conveyors, and other minor items would be depreciated for 5- to 10-year lives under straight line methods; however, the effects of those refinements would be minor.

Investment Tax Credit

Current IRS guidelines allow a tax credit for various types of business equipment. For equipment having a life greater than 7 years, this credit is equal to 10% of the total cost of the equipment. Buildings and land do not qualify for the credit. The entire capital investment (less building) can be considered to have a life of over 7 years for investment tax credit computations.

In addition, wood systems are alternate fuels projects which qualify for an energy investment tax credit of 10% of the total cost. The same capital basis can be used for this computation as for the previous credit.

Limitations applying to tax credits are based on the corporation's total tax returns. It may be assumed that all credits will be allowed with no limitations. Again, a further refinement of these computations should be made prior to financial commitments. Even with some broad assumptions, the computations can become tedious for a 20-year analysis. It is advantageous to utilize a computer to generate annual cash flow tables.

The computer can speed the calculations; however, its power lies in being able to generate summaries of the analysis while varying one or more inputs. This allows the user to see the sensitivity of the project's economics to changes in estimated input(s). Confidence levels in the project are then generated regardless of estimation errors.

Wood system analyses have great sensitivities to fuel prices (as might be expected). Sensitivity to investment capital is usually slight, and to operating costs negligible.

The following is a brief summary of one analysis program which has been successfully utilized for wood combustion systems. Program inputs and outputs are shown in the following outline.

I. *DATA INPUTS*
    A. *Capital Investment*
       Inputs are taken for:
       1. Total capital of project
       2. Equity (amount not financed)
       3. Amount qualifying for tax credit
       4. Amount to be depreciated
    B. *Depreciation*
       One of these selections can be made for depreciation:
       1. Straight line calculation method
       2. Declining balance method
       3. Sum of the years digits method

If declining balance is asked for, an input is made for the acceleration rate (i.e., 200%).

C. *Rate Inputs*

Various rates can be selected as follows:
1. Tax credit rate
2. Interest rate on borrowed capital
3. Rate for inflation (discount rate)
4. Operating cost escalation rate
5. Fuel cost escalation rate
6. Revenue escalation rate (2 inputs are allowed, one for years 1–10, one for years 11–N)

Other Inputs
1. Life of the analysis (years)
2. Title of the project
3. Selection of whether returns and paybacks are to be calculated on a total capital or equity only basis. (This is useful if a project has some government funding)
4. Selection of one variable for a sensitivity analysis (if a sensitivity analysis is desired)

Eligible variables include:
1. Capital
2. Interest rate
3. Life of project
4. Fuel cost (wood price)
5. Fuel consumption (wood consumption)
6. Current fuel cost (natural gas price and consumption)
7. Current fuel cost (oil prices and consumption)
8. Fuel escalation rate (wood escalation rate)
9. Current fuel esc. rate, year 1–10
10. Current fuel esc. rate, year 11–N

If a variable is selected, an input is made for how many entries will be made for the variable and for those entries (i.e., 3 inputs for wood price; $8.00, $10.00, $12.00)

II. *OUTPUTS*

Two main outputs are available:
1. The year-by-year life cycle cash flow table
2. The sensitivity table

The life cycle analysis is prefaced by a listing of all input parameters and selections. Each year has cash flow columns designated as:
1. Year
2. Payment on borrowed capital

3. Operating cost (additional cost for wood systems)
4. Fuel cost (wood cost, annual)
5. Total cost
6. Current annual fuel cost
7. Gross savings
8. Depreciation
9. Tax
10. Net income after taxes
11. Present value (discounted cash flow)
12. Net present value (summation of discounted cash flow)

Final outputs are:
1. Discounted cash flow return on investment
2. Simple payback
3. Discounted payback

Capital expenditure and tax credit is assigned in year "0" (assumed to be the construction period with no operational income or costs)

Program outputs are shown in Figures 9-1 and 9-2. These figures are for a wood system having the following characteristics:

1. Total Capital Cost	$700,000
a. Capital is 100% financed; terms:	20%/20 years
b. Total capital qualifying for tax credit	$637,500
c. Total capital is assumed depreciable	$700,000
2. Additional Operating Costs (1st year)	$20,600
a. Escalation rate per year	7%
3. Wood Fuel Cost (1st Year)	$96,800
a. Wood cost	$8.00 per ton
b. Wood consumption per year	12,100 tons/year
c. Escalation rate per year	9%
4. Current Fuel Cost (1st Year)	$360,163
a. Natural gas cost	$4.56/MCF
b. Natural gas consumption per year	54,850 MCF
c. Oil cost	$.79/gal
d. Oil consumption per year	139,000 gals
e. Escalation rate per year	15% 1st 10 yrs
	10% 2nd 10 yrs

```
CAPITAL INVESTMENT= 700000.
CAPITAL EQUITY =        0
TAX CREDIT CAPITAL = 637500.
TAX CREDIT RATE =20.%
DEPRECIABLE CAPITAL= 700000.
INCOME TAX RATE=  52.%
INTEREST RATE=  20.%
LIFE= 20 YEARS
FIRST YEAR ADD'L OPERATING COST=  20600.
WOOD FUEL COST =     8.00 PER TON
WOOD FUEL CONSUMPTION (YEAR 1)=  12100.TONS PER YEAR
PRICE OF NATURAL GAS = 4.56 PER MCF
NATURAL GAS CONSUMPTION (YEAR 1) =  54850. MCF
PRICE OF OIL =  .79 PER GAL.
OIL CONSUMPTION = 139300. GALS.
RATE FOR NET PRESENT VALUE= 20.0%
OPERATING COST ESCALATION RATE=  7.%
WOOD COST ESCALATION RATE=  9.%
CURRENT FUEL ESCALATION RATE, YEARS 1-10,= 15.%
CURRENT FUEL ESCALATION RATE, YEARS 11-N,= 10.%
```

*DEPRECIATION IS CALCULATED BY
THE 200.% DECLINING BALANCE METHOD

YEAR	CAPITAL PAYMENT	ADD'L OPERATE COST	WOOD FUEL COST	TOTAL COST	CURRENT FUEL COST	GROSS SAVINGS	*DEPRE- CIATION	TAX	CASH FLOW	PRESENT VALUE	NET PRESENT VALUE
0	0							-127500.	127500.	127500.	127500.
1	143750.	20600.	96800.	261150.	360163.	99013.	70000.	15087.	83926.	69939.	197439.
2	143750.	22042.	105512.	271304.	414187.	142884.	63000.	41540.	101344.	70378.	267817.
3	143750.	23585.	115008.	282343.	476316.	193973.	56700.	71382.	122591.	70944.	338761.
4	143750.	25236.	125359.	294344.	547763.	253419.	51030.	105242.	148177.	71459.	410219.
5	143750.	27002.	136641.	307393.	629927.	322534.	45927.	143836.	178698.	71815.	482034.
6	143750.	28893.	148939.	321581.	724416.	402836.	41334.	187981.	214855.	71954.	553989.
7	143750.	30915.	162343.	337008.	833079.	496071.	37201.	238612.	257459.	71852.	625840.
8	143750.	33079.	176954.	353783.	958041.	604258.	33481.	296804.	307454.	71504.	697344.
9	143750.	35395.	192880.	372024.	1101747.	729723.	30133.	363787.	365936.	70921.	768265.
10	143750.	37872.	210239.	391861.	1267009.	875148.	27119.	440975.	434173.	70121.	838387.
11	143750.	40523.	229161.	413434.	1393710.	980276.	27119.	495641.	484635.	65226.	903613.
12	143750.	43360.	249785.	436895.	1533081.	1096186.	27119.	555915.	540271.	60595.	964208.
13	143750.	46395.	272266.	462411.	1686389.	1223978.	27119.	622367.	601612.	56229.	1020437.
14	143750.	49643.	296770.	490162.	1855028.	1364865.	27119.	695628.	669238.	52125.	1072561.
15	143750.	53118.	323479.	520347.	2040530.	1520184.	27119.	776394.	743790.	48276.	1120837.
16	143750.	56836.	352592.	553178.	2244584.	1691406.	27119.	865429.	825977.	44675.	1165513.
17	143750.	60815.	384326.	588890.	2469042.	1880152.	27119.	963577.	916575.	41313.	1206826.
18	143750.	65072.	418915.	627736.	2715946.	2088210.	27119.	1071767.	1016443.	38179.	1245004.
19	143750.	69627.	456617.	669993.	2987541.	2317547.	27119.	1191022.	1126525.	35261.	1280265.
20	143750.	74500.	497713.	715963.	3286295.	2570332.	27119.	1322470.	1247861.	32549.	1312815.

D.C.F. RETURN ON INVESTMENT= 27.8 %

SIMPLE PAYBACK= 4.7 YEARS

DISCOUNTED CASH FLOW R.O.I., SIMPLE PAYBACK,AND
DISCOUNTED PAYBACK ARE COMPUTED ON TOTAL CAPITAL

Figure 9-1.

```
SENSITIVITY ANALYSIS

  1000 h.p. system
  CAPITAL INVESTMENT = $700,000.

  NATURAL GAS PRICE =   4.56 PER MCF
  OIL PRICE =    .79 PER GAL.


        WOOD COST                 D.C.F.R.O.I.     S.P.     D.P.

           8.00                      27.8          4.7      8.0
          10.00                      26.4          5.0      8.9
          12.00                      25.0          5.4      9.9
          15.00                      22.8          6.2     11.8
          18.00                      20.6          7.0     14.5
          20.00                      19.1          7.6     17.1



  HOW 'BOUT THEM CHIPS!!!
```

Figure 9-2.

A sensitivity analysis was performed for wood fuel cost. A range of prices was selected ($8.00 to $20.00) and a summary was made for five discreet prices: $8.00, $10.00, $12.00, $15.00, and $20.00 per ton. In this case, even the highest price of $20.00 per ton showed the project still profitable.

<h1 style="text-align:center">10</h1>

<h1 style="text-align:center">Wood Fuel Processing Routes and Economics</h1>

INTRODUCTION

The nonforest products industry is confronted with a wide variety of options when it decides to switch to wood fuel. Should wood residue be purchased as fuel, or are whole tree chips preferred? What type of boiler and burner are indicated? Is drying or pelletizing wood economically attractive? This chapter reports on work done to quantify the total cost to the user for each wood fuel processing route as the cost of energy (typically as steam) in dollars per million Btu's delivered. (This work was originally reported in June 1980 in reference 40, and for a more complete treatment, this reference is recommended.)

The major emphasis today is placed on commercialized processes such as direct combustion of wood in boilers. However, state-of-the art topics such as wood densification and gasification are approaching commercialization. These processes (which are still in the experimental stage) are dealt with briefly to assess market areas and provide rough cost estimates.

ECONOMIC ANALYSIS

The bottom line for any fuel system is the total of capital, operating, and fuel costs. Each company uses a different method to decide on the viability of an investment—dictated by their method of financing, tax liabilities, and so on. It is difficult to compare fossil-fueled systems using gas or oil with solid-fueled systems using wood or coal. The capital investment and fuel costs are radically different. For these reasons, uniform annual cost and payback period are used in this chapter. The uniform annual cost is divided by the useful heat delivered in order to allow comparison of options over a range of size.

Net present worth, another widely used method, is not presented, but can be developed using the costs contained in this chapter. Net present worth has a weakness in analyzing fuel systems, in that fuel costs for the proposed system and its alternate must be projected over the lifetime of the equipment. The rate of inflation must be factored in for maintenance, taxes, etc., all of which are difficult to predict in the near future and impossible to predict 20 years from now.

Economic decisions are made, however, and a frequently used rule of thumb is a three-year payback period. This varies from 2 to 5 years depending on the market and type of industry.

The technique for calculation of system cost is the same for all systems. The total capital required for the system is computed by adding the cost of the components. An investment tax credit of 20% for wood systems and 10% for coal, gas, and oil systems is utilized. The difference between the total capital cost and the investment tax credit is the financed capital cost—the amount of money which the user must borrow. The annual cost of capital is based on a 15% interest rate and a loan period of 25 years. The maintenance cost, tax, and insurance are based on a fixed percentage of the total capital cost developed from existing installations. The operating costs are estimated from existing installations and vary with size and type of system. The fuel cost is computed using the fuel consumption rates and the unit cost of the various fuels.

PROCESSING ROUTES

Wood fuel can be used in a variety of ways. The simplest and most widely used processing route is the direct combustion of green (50% moisture content) wood chips for producing steam. This route has a minimum of intermediate steps, composed of unloading, storage, conveying, and burning. More complex routes include sizing, drying, densification, gasification, and liquefaction before combustion. Direct combustion is the most viable route, and will be dealt with in detail. Other operations which show future promise are presented in a condensed form. In Chapter 11, a complete graphic of wood processing routes is presented.

WOOD-DERIVED FUELS

The transformation of wood into a more usable form is technically feasible and rapidly approaching economic viability. Included in this category are gasification, liquefaction, and densification. These operations upgrade the form of the fuel by cutting the weight and volume and reducing the cost of transportation and combustion.

Wood gasification is under development in more than 30 companies and institutions in the United States, Canada, and Europe. Gasifiers fueled with wood and coal were used from the late 1800s through World War II when cheap oil and gas forced them from the market place. Their prime market at this time is the retrofit conversion of gas/oil boilers, dryers, and kilns. They can also be used to fuel internal combustion engines or turbines. Gasifiers produce a low-Btu gas which is used on-site, and typically contains 150 Btu/ft^3.

Wood can be used as feedstock for producing ethanol, methanol, pyrolysis oil, and catalytic oil. Ethanol is made by converting cellulose into glucose followed by fermentation. The methanol route uses wood gasification to produce a CO/H_2 mixture followed by shift reaction. Pyrolysis oil, char, and gas are produced by thermal decomposition in the absence of air. Catalytic oil is made using alkaline catalysts at high pressures and temperatures. The alcohols can be used for transportation and may command higher prices, while the oils would be burned in place of fuel oils.

Densified wood is being produced in the form of wood pellets. This process includes drying, grinding, and producing pellets by forcing the finely ground wood through a die. This fuel can be substituted for coal in existing equipment and may find markets in gasifiers and commercial installations.

The results of the economic analysis of wood-derived liquid and gas fuels is found in Table 10-1. The total cost of each product is based on capital, operating, and fuel costs for typical plants. (A more detailed analysis may be found in reference 40.)

The cost of the liquid fuels is higher than competing fossil fuels, but the gap is closing rapidly as gasoline and fuel oil prices increase. Wood pellets, more costly than raw wood or coal, compete with natural gas. Low Btu gas is cost competitive with natural gas and less costly than oil; hence, the renewed interest in this technology.

TABLE 10-1. Cost to Produce Wood-Derived Liquid and Gas Fuels.

FUEL	COST/UNIT	COST/MILLION BTU'S
Ethanol	$1.31/gal	$15.50
Methanol	$0.84/gal	$13.00
Pyrolysis Oil	$0.60/gal	$ 5.17
Pyrolysis Char	$114/ton	$ 4.32
Catalytic Oil	$1.35/gal	$ 9.10
Low Btu Gas	$0.39/MCF	$ 2.57
Wood Pellets	$45/ton	$ 2.94

DIRECT COMBUSTION FOR STEAM GENERATION

Wood presently supplies approximately 2% of the U.S. energy needs, with many projections showing a possible contribution of 7% possible on a renewable basis. The majority of present use is in producing steam for industrial process heat in pulp and paper mills. Nonforest product industries located in the forested areas of the U.S. are now turning to this resource. Typical new installations are wood-fired boilers of 50,000 lb/hr steam capacity rated at 150 psig saturated steam.

The capacities of individual boilers and entire boiler plants span the ranges of 10,000 to 200,000 lb/hr of steam. Five different boiler sizes are analyzed in this range, using seven different systems. These systems are a wood waste boiler, wood chip boiler, wood pellet boiler, wood fueled fluidized-bed combustor, coal boiler, gas boiler with oil backup, and an oil boiler. The costs of the wood waste gasifier with gas/oil boiler are presented in the summary tables and charts. Five different sizes are analyzed for each of these systems: 10,000 lb/hr steam, 25,000 lb/hr steam, 50,000 lb/hr steam, 100,000 lb/hr steam, and 200,000 lb/hr steam. The systems have been configured for almost completely automatic operation, although some applications may not require the degree of automation placed here.

The technique used to analyze each of the seven systems is the same. For each system, a process diagram is developed which shows the routing of the fuel from arrival at the plant to combustion. The cost of each piece of equipment is identified and entered in a system cost table.

The object of the system cost table is to determine a cost of steam in dollars per million Btu's delivered for each system analyzed. These costs

TABLE 10-2. Design Factors.

BOILER OPERATING CONDITIONS

Operating Pressure	= 150 psig saturated
Inlet Feedwater Temperature	= 220°F

EFFICIENCIES (%)

Waste Wood Boiler	= 65	Wood Gasifier	= 85
		(with gas boiler)	= 69
Wood Chip Boiler	= 65	Coal Boiler	= 83
Wood Pellet Boiler	= 80	Gas Boiler	= 81
Wood Fueled Fluidized Bed Combustor	= 68	Oil Boiler	= 85

FUEL DATA

FUEL	HEAT VALUE	UNIT COST	RAW FUEL COST ($ PER MILLION BTU)
Waste Wood	4,250 Btu/lb	$ 7.00/ton	0.82
Wood Chips	4,250 Btu/lb	$10.50/ton	1.24
Wood Pellets	7,650 Btu/lb	$45.00/ton	2.94
Coal	13,000 Btu/lb	$35.00/ton	1.35
Gas	1,000 Btu/gal	$0.30/$10^5$ Btu	3.00
#6 Oil	150,000 Btu/gal	$0.65/gal	4.33

PLANT OPERATION

Hours/day	= 24
Days/year	= 345
Load Factor	= 70%

are then compared to determine the attractiveness of the various systems. The cost of steam is computed by dividing the total annual cost of a system by the annual amount of heat delivered.

The design factors used in the analysis of each of the systems are presented in Table 10-2. Each of the boilers is assumed to generate 150 psig saturated steam, with inlet feedwater at 220°F and efficiencies as shown. The fuel data presented is considered to be representative of current heat values and costs in Georgia. The plants analyzed are assumed to operate 24 hours per day, 345 days per year, and at an average load factor of 70%. This data can be modified as required for analysis of a particular installation.

Waste Wood Boiler System

The process diagram of the waste wood boiler system is shown in Figure 10 1. The waste wood is delivered to the plant by truck, unloaded

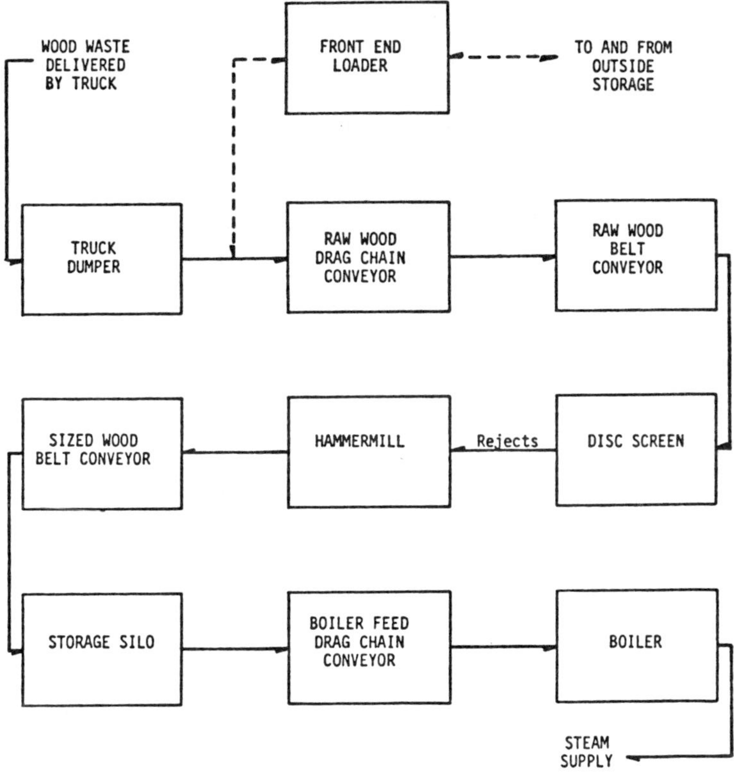

Figure 10-1. Process diagram waste wood boiler system.

with a hydraulic truck dumper, and either immediately processed into the system or moved to and from outside storage using a front end loader. The wood enters the automatic system through a drag chain conveyor which feeds a belt conveyor having a 20-foot horizontal section for fuel transfer. The wood waste is processed over a disc screen where the rejects are further reduced in size by a hammermill. The fines are allowed to pass through without size reduction. Whole tree chips, on the other hand, do not follow this processing route. The sized wood is transported by belt conveyor to a storage silo with sufficient capacity for 3 days of boiler operation. The fuel from the storage silo is transported to the boiler using a drag chain conveyor.

The system cost for the waste wood boiler system is shown in Table 10-3. Since the system utilize wood, a 20% investment tax credit is

TABLE 10-3. System Costs Wood Waste Boiler System.

CAPACITY (LBS/HR)	10,000	25,000	50,000	100,000	200,000
Capital Costs					
Truck Dumper	120,000	120,000	120,000	240,000	240,000
Front End Loader	16,000	22,000	57,000	120,000	240,000
Raw Wood Drag Chain Conveyor	18,000	25,000	38,000	65,000	116,000
Raw Wood Belt Conveyor	56,000	58,000	60,000	67,000	74,000
Disc Screen	7,000	13,000	23,000	37,000	66,000
Hammermill	22,000	30,000	43,000	71,000	125,000
Sized Wood Belt Conveyor	154,000	163,000	172,000	189,000	200,000
Storage Silo	32,000	79,000	158,000	315,000	630,000
Boiler Feed Conveyor	41,000	43,000	47,000	53,000	68,000
Boiler	350,000	625,000	900,000	2,800,000	4,800,000
Total Capital Cost	816,000	1,178,000	1,618,000	3,957,000	6,319,000
Less Investment Tax Credit	163,000	236,000	324,000	791,000	1,264,000
Financed Capital Cost	653,000	942,000	1,294,000	3,166,000	5,055,000
Annual Cost of Capital	101,000	146,000	200,000	490,000	782,000
Operating Cost	134,000	150,000	194,000	352,000	379,000
Maintenance Cost	41,000	59,000	81,000	198,000	316,000
Tax and Insurance	20,000	29,000	40,000	99,000	158,000
Fuel Cost	74,000	184,000	370,000	741,000	1,480,000
Total Annual cost	370,000	568,000	885,000	1,880,000	3,115,000
Annual Heat Delivered (10^6 Btu)	59,000	146,000	292,000	585,000	1,168,000
Cost Per Million Btu's Delivered	$6.27	$3.89	$3.03	$3.21	$2.67

used. The operating cost is a percentage of the total capital cost—taken from operating experience. The maintenance cost is assumed to be 5% of the total capital cost, and the tax and insurance are assumed to be 2.5% of the total capital cost. The annual heat delivered is calculated by determining the amount of heat released for a 24-hour day operation, 345 days per year, at a 70% load factor.

Wood Chip Boiler System

The process diagram for the wood chip boiler system is similar to Figure 10-1; however, the disc screen, hammermill, and raw wood belt conveyor are eliminated. Since the wood chips already have a somewhat uniform size, no size reduction equipment is required, and the fuel is transported immediately to a storage silo with a capacity for 3 days boiler operation. Wood is transported from the storage silo to the boiler using a drag chain conveyor.

The costs for this system and the other systems which follow were computed in a similar manner to those for the waste wood system. Total fuel and capital costs are presented in Figures 10-2 and 10-3.

Wood Pellet Boiler System

Wood pellets are dumped directly into a bin which feeds a belt conveyor. An advantage of pellets is their low moisture content. Since it is not desirable to add moisture to the pellets, no outside storage is utilized. A drag chain conveyor would tend to break the pellets, therefore, a belt conveyor is used to feed the storage silo and then to feed the boiler.

Wood Fueled Fluidized-Bed Combustor System

Fuel for the system (waste wood) is delivered to the plant by truck, unloaded by a hydraulic truck dumper, and transported to and from outside storage by a front end loader. The wood enters the system through a drag chain conveyor which feeds a belt conveyor. A disc screen is again utilized to separate oversized pieces of wood, and a hammermill is used for size reduction. A belt conveyor transports the wood to a storage silo, and a drag chain or screw conveyor is used to transport the wood from the storage silo to the fluidized-bed combustor.

Coal Boiler System

Coal is delivered to the plant by truck, unloaded by a hydraulic truck dumper, and moved to and from outside storage with a front end loader. The coal enters the combustion system through a drag chain conveyor feeding a belt conveyor. It is assumed that the coal is already sized— thus no size reduction equipment is required. A belt conveyor will feed a storage silo, and the coal is then transported from the storage silo to the boiler with a drag chain conveyor.

While a hydraulic truck dump is specified for unloading, many plants would choose rail delivery of coal. A short spur line with bottom unloading via a trestle would cost approximately $235,000. Deleting the cost of the truck dump, this system would add $.03 per million Btu's delivered. Long spur lines might increase this cost considerably, and the

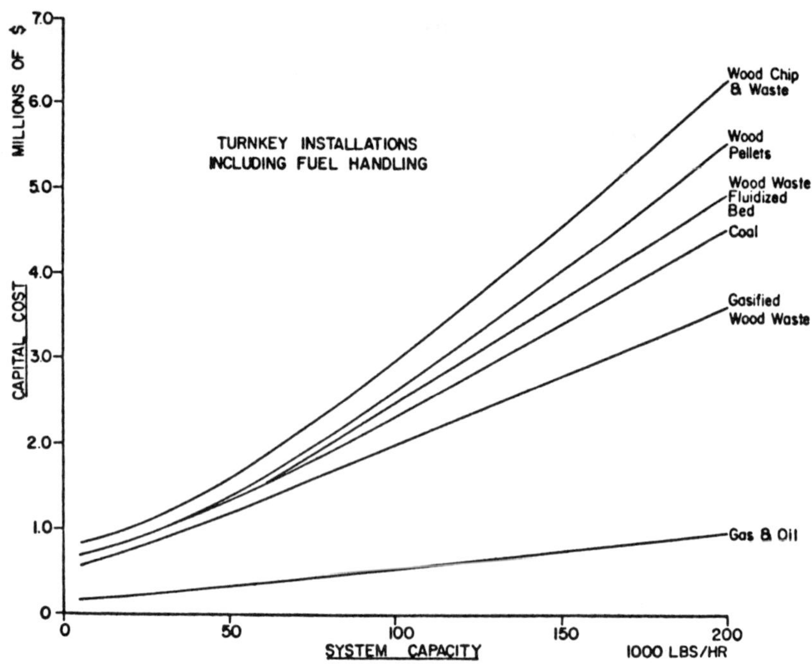

Figure 10-2. Capital cost of steam generating system.

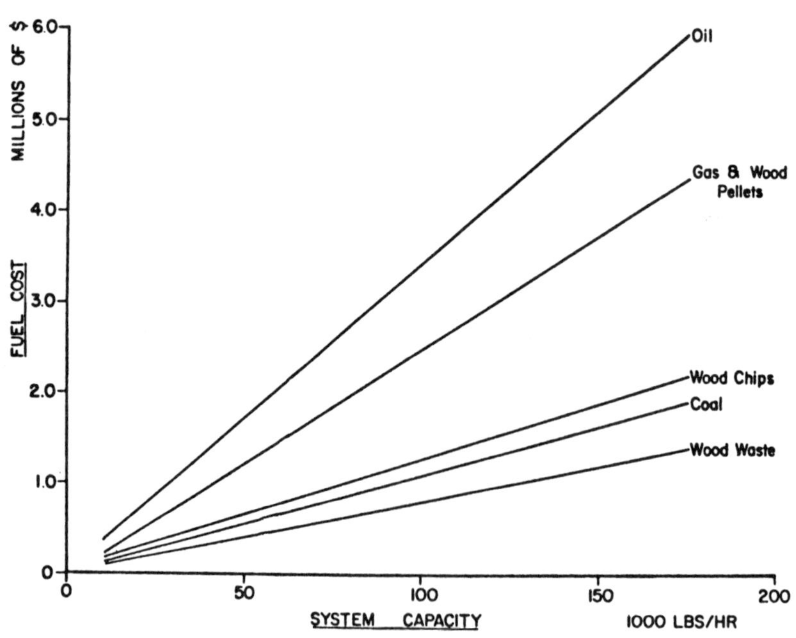

Figure 10-3. Annual fuel cost for steam generating systems.

choice of rail or truck delivery would depend on availability of rail service, size of boiler, and distance to coal mines.

The capital and operating costs of SO_2 scrubbing equipment is not included. Relatively low sulfur coal (less than 2% sulfur) is available from Tennessee and Alabama, and SO_2 clean up is usually not required for boilers of up to 200,000 lbs/hr steam capacity. The addition of SO_2 scrubbers would add approximately $5 per lb/hr of steam capacity to the capital cost of the boiler plant. The annual cost for this capital, plus maintenance, would add approximately $.25 per million Btu delivered to the cost of a coal-fired steam system.

Gas Boiler System

Due to the possibility of a curtailment of natural gas supplies to an industrial boiler, the system is designed to fire either gas or oil. The gas is supplied to the boiler directly from a pipeline; oil is delivered to the plant by truck. An outside oil storage facility is equipped with a transfer pump, which pumps the oil to a day storage tank which in turn feeds the boiler.

The outside oil storage tank is sized for a 7-day emergency supply of oil; its pumps are designed to fill the day tank in one hour. The day tank and outside storage tank are equipped with heaters for the #6 fuel oil.

Oil Boiler System

Oil is delivered to the plant by truck, stored in an outside storage tank, and pumped by transfer pump to the day tank. The day tank transfers oil directly to the boiler with pumps.

The outside oil storage tank is designed for a capacity providing 30 days of boiler operation. The oil transfer pumps are designed with a capacity to transfer oil to the day tank in one hour, both day tank and outside storage tank have heaters for the #6 fuel oil.

Wood Waste Gasifier with Gas/Oil Boiler

An updraft gasifier used to retrofit a gas/oil boiler is the basis for this system. It is fueled with whole tree chips and is close-coupled to the boiler. The total capital costs, fuel costs, and steam costs are presented in the summary table for comparison with other systems.

SUMMARY

Capital Costs

The capital cost of any steam generating system can be separated into *boiler cost* and *fuel system cost.* The latter can be found by subtracting the boiler cost from the total cost. Comparison of the cost data and the efficiency data indicates that boilers for gas and oil systems are efficient and low in cost, and fuel system costs are minimal. In contrast, solid fuel equipment is less efficient, particularly when burning high moisture content fuels. The boiler cost and extensive solids handling and storage systems require a large capital investment. Solid fuel equipment is also considerably larger in size.

Figure 10-2 presents capital cost curves for boiler systems including fuel handling from 10,000 to 200,000 lb/hr steam for gas, liquid, and solid fuels. Solid-fueled boiler systems are approximately 9 times the cost of gas/oil systems. Gasifier retrofit of existing boilers is somewhat less, at approximately 5 times the cost of gas/oil systems.

Fuel Costs

Figure 10-3 summarizes the fuel cost for the same systems based on a 345-day-per-year operation and a 70% load factor. Oil is the most costly fuel (No. 6 @ $.65 per gallon) followed by natural gas, wood pellets, wood chips, coal (less than 2% sulfur), and wood waste. It is interesting to note that an oil-fired boiler of 30,000 lb/hr capacity with an initial cost of $210,000 will consume $1 million worth of fuel oil in its first year of operation. The annual fuel cost for most solid fuel systems is about ⅓ of that for gas or oil systems.

The capital cost and fuel costs of the eight steam production systems are inversely proportional; i.e., the cheapest fuel requires the most costly handling and burning systems.

Steam Costs

The primary application in industry for wood fuel and competing fossil fuels is steam production. Using the costs for raw fuel and plant investment covered previously and adding operation costs as a percentage of plant investment, the total cost of producing steam can be found. This cost can be used to compare the economic merits of wood- and fossil-fueled systems for boiler replacement or plant expansion.

The cost of steam for the individual systems is summarized in Table 10-4 and Figure 10-4. Cost is presented in dollars per million Btu's

TABLE 10-4. Cost of Steam (dollars per million Btu's delivered).

	SYSTEM RATING				
	10,000	25,000	50,000	100,000	200,000
Wood Waste Boiler	6.27	3.89	3.03	3.21	2.67
Wood Chip Boiler	6.37	4.29	3.53	3.76	3.30
Wood Pellet Boiler	7.96	5.91	5.10	5.29	4.90
Wood Waste Fluidized-Bed Combustion	7.86	4.61	3.45	3.06	2.68
Coal Boiler	5.76	4.05	3.44	2.98	2.69
Gas/Oil Boiler	4.61	4.34	4.16	4.05	3.99
Oil Boiler	6.12	5.80	5.61	5.45	5.38
Wood Waste Gasifier on Gas/Oil Boiler	4.54	3.10	2.50	2.22	2.01

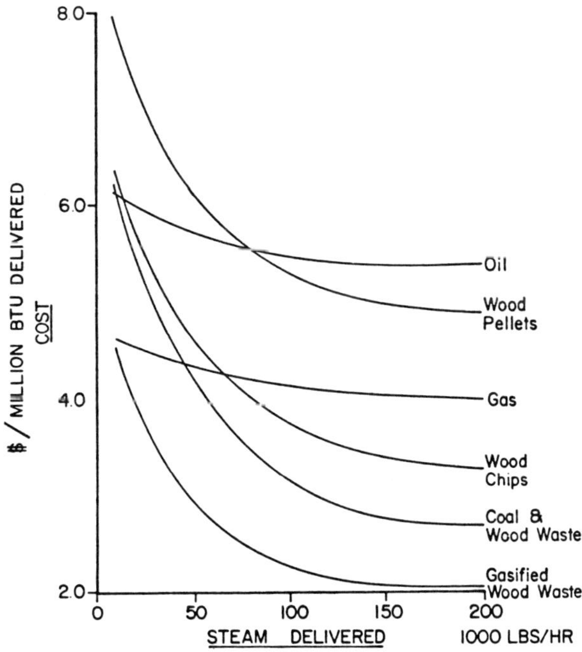

Figure 10-4. Total cost of steam.

delivered of 150 psig saturated steam. This data indicates that the cost of steam goes down with increasing boiler size for all fuels, with the minor exception of a slight jump in price from package boilers to the larger field-erected type over 50,000 lb/hr steam capacity. For the smallest boilers (10,000 lb/hr capacity), natural gas is the least costly system. However, interruptions and rising gas prices may change this position in the future. Coal is more costly, followed by oil, wood waste, wood chips, wood waste for a fluidized-bed combustor, and wood pellet boiler in order of increasing cost. This analysis points out the fact that solid fuels are expensive to use for small industrial applications.

The 25,000 lb/hr size range is perhaps more indicative of industrial boiler sizes. In this size, wood waste is the lowest cost, followed by coal, wood chips, and gas. Natural gas is more costly, followed by wood pellets and fuel oil.

The cost of wood gasifiers retrofitted to existing gas/oil boilers is of interest, although the economic analysis is complicated by the "existing" boiler which may already be fully depreciated. In order to keep the analysis on an equal basis, the cost for delivered steam from a gasifier retrofit system includes the amortization, operation, and maintenance costs of a gas boiler. These costs, based on waste wood, are the least of any system in any size range. Using wood chips as fuel, the cost per million Btu delivered as steam increases $.61, making it the least costly option—with the sole exception of the natural gas boiler at 10,000 lb/hr capacity. Ths economic advantage is countered by the lack of dependable hardware at present. Over 30 groups in the U.S. and Canada are developing wood gasifiers and availability problems may be alleviated in the next few years.

The economics of firing boilers with dry, densified wood pellets are competitive only with fuel oil in the 50,000 lb/hr size range and above. Cost reductions in pelletization technology or the emergence of compact, low cost gasifiers to convert pellets into a gas compatible with existing gas/oil equipment should greatly improve this competitive position.

The solid fuel systems with larger capital costs will be more sensitive to increases in loan interest rates than oil and gas systems. For example, changing from a 15 to 17% interest rate will increase the cost of steam for a 25,000 lb/hr wood waste boiler by $0.11 per million Btu but only $0.03 per million Btu for a 25,000 lb/hr gas/oil boiler.

Table 10-5 shows the payback period for wood waste, chip, and pel-

TABLE 10-5. Payback Period for Steam Generating Systems (compared to conventional oil boiler).

CAPACITY (LB/HR)	PAYBACK PERIOD (YEARS)		
	WASTE	CHIPS	PELLETS
10,000	Oil cheaper	Oil cheaper	Oil cheaper
25,000	3.52	4.00	Oil cheaper
50,000	1.77	1.99	6.90
100,000	2.63	3.30	27.90*
200,000	1.69	2.20	8.20

*Due to rising capital cost for field erected boiler.

let fuel systems as compared to a conventional oil-fired boiler. The cost data computed indicates that oil is a cheaper system in the 10,000 lb/hr size range. The cost of the wood systems could be reduced by removing some of the automation and storage designed into the system. Wood waste and chips will payoff within the five-year period which is generally considered to be the limit for industrial applications. Pellets are not competitive in the smaller capacities and have unacceptably long payback periods in the larger sizes. As previously noted, pellets are not viable fuel when analyzed by other economic methods. The jump between 50,000 and 100,000 lbs/hr is due to the increased capital cost of a field-erected versus a packaged boiler. As capacities increase above 100,000 lbs/hr, the boiler costs and the resulting payback periods decrease due to economies of scale.

Section III

11

Wood Fuel Processing Network

A number of processing routes are available to the energy user when he considers using wood to produce energy. These multiple paths and options can be confusing to the newcomer in the wood energy field, and without adequate explanation, can give the impression that the use of wood for fuel is fraught with difficulties. To clarify the situation, a Wood Energy Processing Network (Figure 11-1) has been developed.

The network has three major phases: source processing, fuel processing, and site processing. The network shows the various routes involved in each phase of wood energy processing.

For example, the network indicates the steps required to process the raw source (wood) from the source site. Each block indicates a major process involving capital equipment and manpower. Each process has a cost which must be added to the base price of the raw source itself. The cost of waste and chips after the transport process has been completed is the wood price that is quoted for delivered fuel.

The network is designed to serve as a roadmap for showing the relationship of individual processing steps to the overall conversion of wood to energy. Two areas currently under investigation are fuel standards and supply. These areas relate directly to the wood sources shown under source processing. The wood fuel supply and the standards applied to the fuel will affect the cost availability of the basic fuel source. Obviously, a change in this area will affect every part of the network.

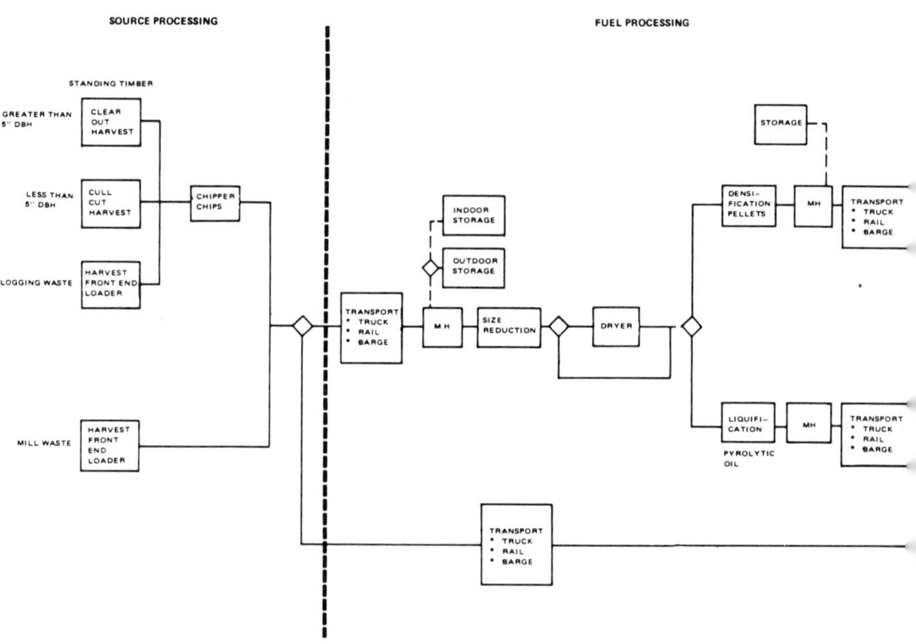

Figure 11-1. Wood energy processing network.

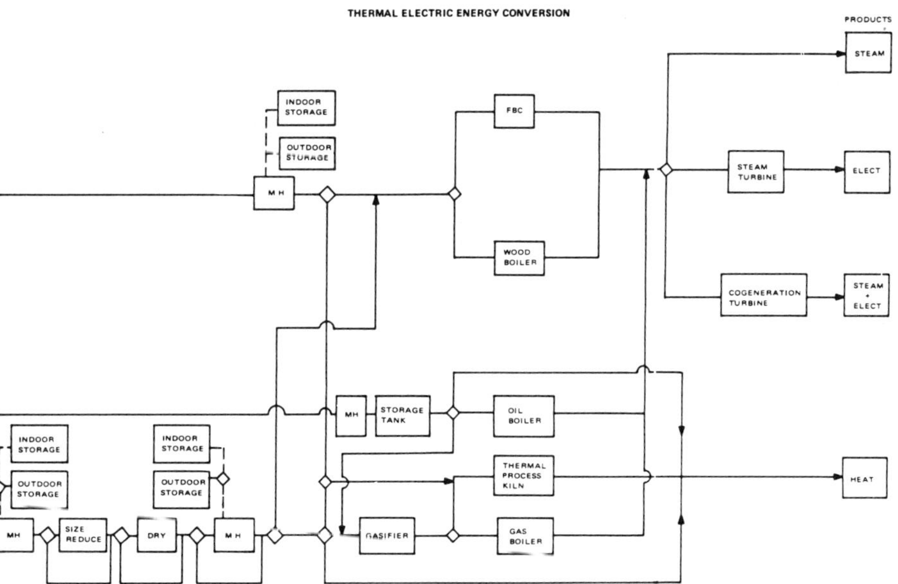

Figure 11-1 (*continued*)

12

Case History—Integrated Products, Inc.

Integrated Products, Inc., is a textile manufacturing firm headquartered in Rome, Georgia. The company manufactures yarn for the carpet industry and has several plants located in Rome, Villa Rica, and Aragon, Georgia. The company employs 1,800 people and supplies products to domestic and international customers. The demonstration project discussed here is being conducted at the Aragon plant located 60 miles northwest of Atlanta. The Aragon plant produces carpet yarn and has approximately 450 employees. The plant uses steam from the boilers to "heat set" and dry the carpet yarn in order to provide a permanent twist to the yarn. This process is accomplished in "auto claves" and steam dryers, and in winter, steam is also used for space heating. Formerly, the plant boilers ran on natural gas and oil; approximately 85% of the energy consumption was provided by natural gas and 15% by No. 2 fuel oil.

Visits to the plant by Georgia Tech personnel and meetings with the vice-president and plant engineers of the company indicated that the Aragon plant could be a prime candidate for a wood energy demonstration project. The potential for success was based on the following considerations:

- A preliminary economic analysis performed on the basis of the energy usage data indicated that a wood energy system would be economically feasible and that the company had a strong potential for conservation of scarce gas and oil fuels.
- Meetings with the engineering and plant personnel revealed that the

company had a competent technical staff interested in the conversion to wood.

- A survey of the local wood supply situation indicated that wood fuel in the form of sawdust, bark, and wood chips would be available at economical prices.
- Financial records of the company revealed that the company was stable and the management had a positive attitude with respect to participating in a demonstration project.
- Integrated Products, Inc., belongs to a nonforest industry, and its success in the demonstration project would serve as a model for other industries to follow.

CONCEPTUAL DESIGN

The primary goals of the conceptual design study of the proposed wood energy system were to determine boiler size, determine the wood fuel storage volume required, determine the optimum wood handling system, and develop plans for the plant layout.

In order to perform this task, information on the energy use of the existing plant, specifications of the existing boilers, characteristics of the steam consuming equipment, and available space for the installation of the wood system was collected. The engineering staff and plant personnel of the company provided a considerable amount of information and assistance in these efforts. Information on the existing boiler plant is summarized in Table 12-1. Monthly energy consumption data is shown in Table 12-2, and a plot plan of the area adjacent to the boiler house is shown in Figure 12-1.

Size of the Boiler

In sizing the boiler, the existing energy consumption, the size and utilization of the existing boilers, characteristics of the steam consuming equipment, and the future plans of the company were taken into consideration. There were no steam flow charts to indicate the total instantaneous demand on the boilers. Steam flow charts connected with the major steam consuming equipment such as the auto claves and steam dryers indicated the cyclical nature of the steam demand by the equipment. Discussions were held with the engineering and plant personnel concerning the loading of the existing boilers and the addition of another

TABLE 12-1. Feasibility Study Background Data.

Name of Company:	Integrated Products, Inc. Aragon, Georgia
Product Line:	"Heat Set" nylon carpet yarn for the carpet industry
Number of Employees:	450
Boiler Plant Information:	There are three boilers with descriptions as follows:

Boiler No. 1
  Date of installation:
  Capacity:
  Pressure:
  Fuel:
  Emission control:
  Ash handling:
  Boiler efficiency:
	"Eclipse" package boiler
	1972
	200 hp (6900 lb steam/hr)
	100 psig
	Gas or oil
	Chimney
	None
	82.5% (approximately)

Boiler No. 2
  Date of installation:
  Capacity:
	"Eclipse" package boiler
	1972
	350 hp (12,000 lb steam/hr)
	Other specifications are the same as those of Boiler No. 1.

Boiler No. 3
  Date of installation:
  Capacity:
  Fuel:
	"Cole" HRT Boiler (used as standby)
	1920
	100 hp
	Gas or oil

Steam Usage:	Steam is used in autoclaves for the "heat set" process. This process provides a permanent twist to the yarn. During winter, approximately 10% of the space heating demand is met by steam.
Plant Operation:	24 hours/day, 6 days/week, 50 weeks/year
Energy Consumption:	Natural gas: $51{,}934 \times 10^3$ cu ft/year No. 2 fuel oil: 53,464 gals/year
Other Information:	There is sufficient yard area available for wood handling, storage, and installation of the wood-fired boiler.

dryer. Based on these considerations, a 400 hp (13,000 lb steam/hr) boiler was judged to be adequate by the engineering and plant personnel of Integrated Products, Inc., and Georgia Tech personnel.

Wood Fuel Storage and Handling

To estimate the size of the storage and the wood handling equipment, the following assumptions were made with regard to wood fuel properties and system efficiency.

Wood Fuel:	Sawdust, bark and chips 50% moisture content
Heat value:	4,000 Btu/lb
Bulk density	24 lb/ft^3
System thermal efficiency	65%
Heat input required to generate 1 lb of steam	1,000 Btu
Calculations:	
Output of boiler at rated capacity	400 hp $\times$ 34.5 lb-steam/hr (boiler hp) $\times$ 1,000 Btu/lb-steam $= 13.8 \times 10^6$ Btu/hr
Wood fuel consumption/hr	13.8×10^6 Btu/hr /4,000 Btu/lb $\times 0.65 = 5,308$ lb/hr
Volume flow rate of wood	221 ft^3/hr
Covered Storage	
Storage volume for 72 hours (rated capacity)	15,912 ft^3

TABLE 12-2. Energy Consumption: Integrated Products, Inc.

PERIOD	MONTH	NATURAL GAS		OIL		TOTAL BTU
		THERMS	BTU	GALS	BTU	
4/28–5/30/78	May 78	49,955	4.9955 x 10^9			4.9955 x 10^9
5/30–6/28/78	June 78	45,093	4.5093 x 10^9			4.5093 x 10^9
6/28–7/28/78	July 78	36,710	3.6710 x 10^9			3.6710 x 10^9
7/28–8/29/78	Aug 78	46,868	4.6868 x 10^9			4.6868 x 10^9
8/29–9/29/78	Sept 78	41,625	4.1625 x 10^9			4.1625 x 10^9
9/29–10/31/78	Oct 78	52,253	5.2253 x 10^9			5.2253 x 10^9
10/31–12/1/78	Nov 78	48,322	4.8322 x 10^9			4.8322 x 10^9
12/1/78–1/3/79	Dec 78	35,782	3.5782 x 10^9	8,000	1.1200 x 10^9	4.6982 x 10^9
1/3/79–1/31/79	Jan 79	27,742	2.7742 x 10^9	23,215	3.2501 x 10^9	6.0243 x 10^9
1/31/79–3/1/79	Feb 79	34,466	3.4466 x 10^9	14,749	2.0649 x 10^9	5.5115 x 10^9
3/1/79–4/1/79	Mar 79	55,409	5.5409 x 10^9	7,500	1.0500 x 10^9	6.5909 x 10^9
4/1/79–4/30/79	April 79	45,123	4.5123 x 10^9			4.5123 x 10^9
ANNUAL CONSUMPTION						59.4198 x 10^9

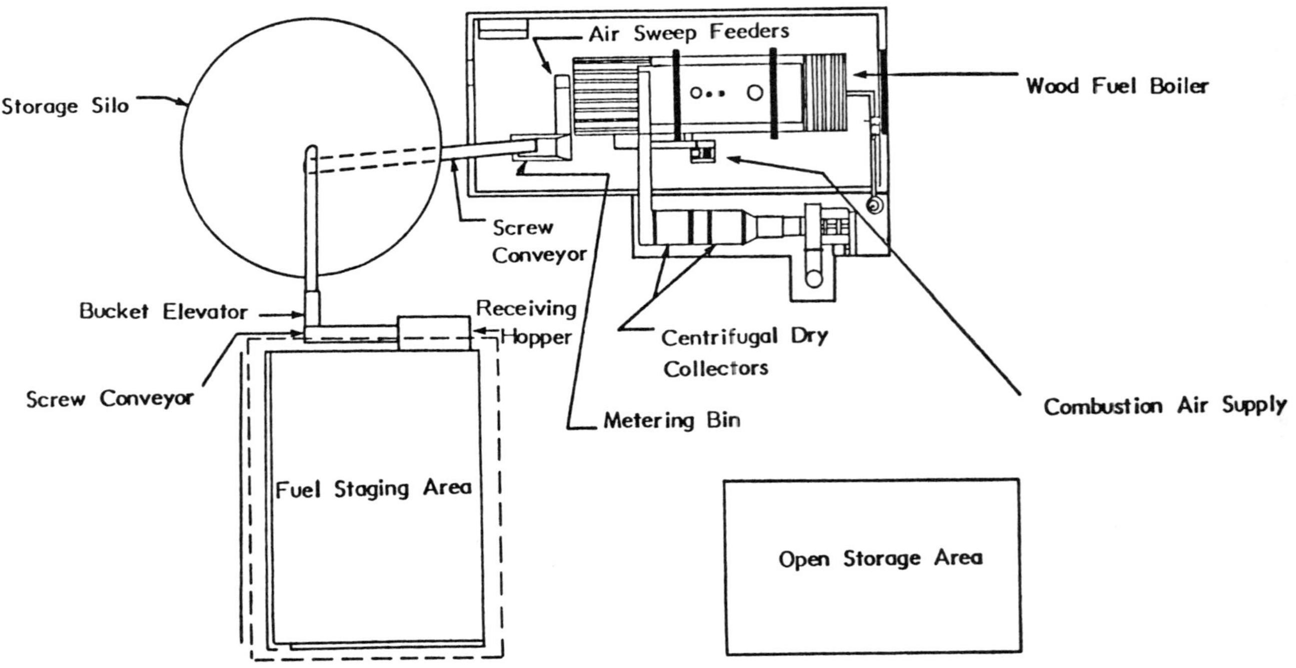

Figure 12-1. Integrated Products Inc., wood-fired boiler plant.

A storage silo of this size or a covered storage shed of floor area of approximately 2,600 ft ^{2} (assuming average of 6 ft of pile depth) can be used.

Open Storage Area
Concrete floor area for 28
   days storage, assuming
   avg. of 12 ft pile depth   12,376 ft ^{2}
Wood Requirements   2.654 tons/hr
Wood required/hour $\simeq$ 3 truckloads/day at rated capacity
                                      $\simeq$ 2 truckloads/day at 60% to 70%
                                        of rated capacity

Wood required/year at
   100% capacity $\simeq$ 19,000 tons/year
Wood required/year at
   70% capacity $\simeq$ 13,300 tons/year

Wood fuel handling.   For a plant of this size, preliminary feasibility studies indicated that an automatic fuel receiving hydraulic truck dump would be an expensive alternative. Simpler systems such as the use of a front end loader or fuel delivery by live bottom delivery vans were considered to be economical alternatives.

Other fuel handling equipment such as conveyors, elevators, etc., designed to handle volume of wood (design speed 221 ft ^{3}/hr) would be included in the system.

Space requirements.   Based on manufacturer's data for wood-fired boilers, wood handling equipment, and our own calculations, it was found that a 400 hp wood system including a storage silo would require approximately 7,200 ft ^{2} of space for the boiler house, silo fuel unloading, and driveway for trucks to enter the fuel staging area.

CONTRACTOR SELECTION

Based on our discussions with the engineering staff of the company and our study of the fuel supply situation, it was decided that the selected boiler should be capable of burning sawdust, bark, wood chips, and other kinds of wood waste up to 50% moisture content on a wet basis. Various boiler types were considered. On the basis of dependability, performance, fuel capability, and first cost, an HRT (Horizontal Return Tube) boiler was considered to be a good choice. A number of boiler manufacturers were contacted based on Georgia Tech experience with previous projects in the wood energy area. The following firms actually visited the plant site:

- Applied Engineering Co.
   Atlanta, Georgia
- Wellons, Inc.
   Sherwood, Oregon
- Kincaid Engineering Company, Inc.
   Gastonia, North Carolina
- Industrial Boiler Co., Inc.
   Thomasville, Georgia
- Energy Systems, Inc.
   Chattanooga, Tennessee

During the visits, the representatives from these firms investigated the details of the existing boiler plant, including space availability and use. Some of the vendors explained the details of the systems they were going to propose including details of similar systems supplied to other companies. Based on their examinations of the plant site and the requirements of the proposed wood energy plant, four of the companies sent their budget estimates as proposals. There were some variations in the components of the systems proposed by these firms. The equivalent costs of the complete systems varied from approximately 460,000 to $600,000. In the selection of the contractor, the following factors were taken into consideration:

- The cost of the wood energy system quoted by the vendor
- The type of wood boiler and the details of the wood energy system
- The reliability and the service that would be furnished by the vendor as determined by previous experience and reputation
- Contractors' guarantees on system performance and emission characteristics

Based on these considerations, the proposal from the Industrial Boiler Company was accepted by the management of Integrated Products, Inc. Before arriving at this decision, the vice president for engineering at

Integrated Products had visited a similar wood energy system in Rome, Georgia, that was built by the Industrial Boiler Company. Georgia Tech personnel also visited this plant.

The decision by the Integrated Products Co. was agreeable to Georgia Tech personnel since the Industrial Boiler Company is a reputable firm and has satisfactorily installed a number of wood energy systems in Georgia and other parts of the United States. All of the functions of design, manufacture, installation, and startup of the entire wood system would be borne by the same contractor. The accepted proposal was the low bid among those submitted by the proposers.

Contract Document Preparation

The proposal submitted by the Industrial Boiler Company included details of the wood energy system and budget prices for the various components. The proposal was reviewed by the engineering staff of Integrated Products Co., and by Georgia Tech personnel with respect to the costs of the system components, the optional equipment, and the system layout. Meetings were also held with representatives from the Industrial Boiler Company and the final contract included the following items:

Fuel Storage System
- Poured concrete silo
- Supreme unloader

Fuel Feeding System (Unloader to Boiler)
- Conveyor—unloader to metering bin
- Metering bin
- Transfer conveyor—metering bin to air sweep feeders
- Air sweep feeders
- Fuel receiving area—sheltered

Steam Generating System
- Industrial horizontal tubular boiler, Model 3-2600-158
- Cast iron pin hole grates
- Combustion air system
- Steel casing
- Refractories
- Rear arch

- Pre-wired automatic control panel
- Primary air pollution control equipment

Exhaust Gas Handling System

- Exhaust gas duct
- Secondary air pollution and control equipment
- Clean gas duct
- Draft inducer (fan)
- Exhaust gas stack

Auxiliary Equipment and Services

- Boiler feed water system
- Chemical feed system
- Steam flow meter
- Piping material for installation of the system within boiler room
- Electrical materials for field wiring of system
- Labor—for installation of system

The boiler house and foundation, the sheltered fuel staging area, and the open storage area were built by a local construction company under a separate purchase order from Integrated Products, Inc.

A purchase order for the system was signed by Integrated Products, Inc., in a meeting on May 19, 1980 at Rome, Georgia. The vice president for sales and chief engineer from Industrial Boiler Company, and Georgia Tech personnel were present at the meeting. (The final system layout is shown in Figure 12-1).

SYSTEM DESCRIPTION

The system (as shown in Figure 12-1) consists of the following major components:

> Horizontal return tube boiler (HRT)
> Wood storage
> Wood fuel handling

Each one of these components is described briefly in the following sections.

Horizontal return tube boiler.   The HRT boiler is a firetube boiler designed to produce 13,320 lbs/hr of saturated steam (400 hp) at 130 psig from feed water supplied to the unit at 220°F. The boiler is designed to burn wood fuel up to 50% moisture content (wet basis) such as sawdust, bark, and chips. Wood fuel enters the grate area by means of air sweep feeders.

Wood storage.   A poured concrete silo, 24 ft diameter x 48 ft high, provided with a supreme unloader, is used for storage. The usable storage capacity of this silo is 14,464 cu ft and is sufficient to store wood fuel for approximately 72 hours, when the boiler is operating at 70% load. For long term storage, an open storage area of approximately 10,000 ft ^{2} provided with a concrete floor and sufficient to store wood for approximately 28 days has been built on adjacent land owned by the company.

Wood Fuel Handling.   Wood fuel is supplied to the plant by means of live bottom trailers. They can unload the fuel in the open storage area or in the fuel staging area which is located close to the wood storage silo. The wood fuel in the fuel staging area is transferred to the receiving hopper located in the corner by means of a front end loader. The receiving hopper is approximately 4 ft wide, 8 ft long, and 6.5 ft high, and the fuel received in this hopper is delivered to the storage silo by means of a screw conveyor and bucket elevator. Fuel from the silo to the metering bin and from the metering bin to the air sweep feeders is transferred by means of screw conveyors.

FEASIBILITY STUDIES

A preliminary economic analysis was performed on the basis of budget estimates for the equipment and wood fuel costs. This was done prior to the acceptance of the Aragon plant as a candidate for the demonstration project. A refined feasibility study based on the quoted prices for the equipment was later prepared. Addressed were major items such as economic analysis and the wood supply situation.

Economic Analysis

The economic analysis includes the costs associated with the wood system and the existing system, the first year savings, and the payback period. Life cycle costs based on escalation rates for the costs and the present worth savings over the life of the equipment were calculated using a computer. The items used in the analysis include:

- Total costs (including operating costs, etc.)
- Total annual savings due to the installation of the wood system
- Approximate payback period
- Life cycle costs and savings allowing for escalation rates

Costs

Costs were based on the information obtained from the manufacturers, plant engineering personnel, and our own experience and discussion with professionals in the field. Current prices of conventional fuel were obtained from Integrated Products.

In computing the capital cost, an allowance for the federal investment and alternative energy property tax credits amounting to 20% of the investment is made. Additionally, the DOE grant to the company toward the demonstration is subtracted to arrive at the final adjusted capital investment.

Cost of the System:	$460,000
Less 20% federal tax credits	$ 92,000
Less DOE Grant	$146,000
Adjusted Capital	$222,000

The system includes the boiler and related equipment described earlier, building, storage, etc.

In arriving at the nonfuel operating costs and fuel costs, the following data and assumptions are used:

Operating Costs
   a. *Wood system:*

Capacity	= 400 hp or 13,800 lbs steam/hr
Wood fuel	= Sawdust, bark, whole tree chips
Moisture content	= 50% (wet basis)
Heat value	= 4,000 Btu/lb

Wood system thermal
   efficiency $= 65\%$

b. *Existing system:*
  Fuel consumption $= 87.4\%$ natural gas
                               12.6% No. 2 oil

  Natural gas $= 51,394,400$ ft^3/year
  No. 2 fuel oil $= 53,464$ gals/year
  Heat value of natural
     gas $= 1,000$ Btu/ft^3
  Heat value of No. 2 oil $= 140,000$ Btu/gal
  System efficiency $= 80\%$ (gas-oil)

c. *Other assumptions:*
  Life of wood system $= 20$ years
  Interest rate $= 12\%$
  Cost of electricity $= \$0.03$/kWh
  Labor $= \$5.62$/hour
  Capital of existing
     system $=$ Fully paid

d. *Fuel costs:*
  Annual energy
     consumption $= 59.42 \times 10^9$ Btu/year (gas & oil)
  Net energy output $= 47.536 \times 10^9$ Btu/year (80% eff)

$$\text{Wood fuel consumption} = 47.536 \times 10^9 \,/\, 4000 \times 2000 \times 0.65 \text{ tons/year}$$
$$= 9142 \text{ tons/year}$$
$$\text{Cost of wood fuel @ } \$8/\text{ton} = 9142 \times 8 = \$73,136/\text{year}$$

Conventional fuel costs
Cost of No. 2 fuel oil $= 53,464$ gallons $\times \$0.86$/gal $= \$ \ 45,979$/year
Cost of natural gas $= 51,934.8$ MCF $\times \$3.00$/MCF $= \$155,804$/year
$$\text{Total cost of conventional fuels} = \$201,783/\text{year}$$

Life Cycle Analysis

The costs above are useful for evaluating the savings for the first year
only (1980–81). In order to assess the economics of this project on a
long term basis, the effects of specific cost increases and inflation have
to be taken into consideration. Therefore, life cycle costs were developed by using escalation rates for the various costs. The following annual escalation rates were used for the various costs:

Operating costs (wood system)	7%
Operating costs (conventional system)	7% per year
Property tax & insurance	6% per year
Wood fuel	9% per year
Gas and oil	17% per year for the first 10 years
	10% per year for the second 10 years
Coal	9% per year

Table 12-3 shows the various costs, the annual savings, and the present worth of the savings for each year over the life of the equipment.

WOOD FUEL AVAILABILITY

As part of the feasibility study, inquiries were made regarding wood fuel availability near the plant. Wood fuel suppliers were contacted both by Georgia Tech personnel and by the staff of the Aragon plant. Wood fuel suppliers located near the plant have shown a great interest in developing a steady wood supply. These potential suppliers include:

- S.L. Miller Mills
    Crystal Springs, Georgia
- Jack Gresham Corporation
    Crystal Springs, Georgia
- Robert House
    Rome, Georgia
- Joe Smith Trucking Company
    Cumming, Georgia

Additionally, the project on wood fuel sources and suppliers shows a study conducted by the University of Georgia as part of the feasibility studies for this project which resulted in a list of suppliers in the counties of Floyd, Paulding, Polk, Carroll, Bartow, and others close to the Aragon plant. This information was made available to the participant.

OPERATIONAL REPORT

As of this writing, Integrated Products' wood-fired system at Aragon has been on line approximately 6 months. A minor problem in the ma-

TABLE 12-3.

NAME OF COMPANY: INTEGRATED PRODUCTS INC.
LIFE CYCLE COSTS:

| | WOOD SYSTEM | | | | | EXISTING SYSTEM | | | | | |
YEAR	ANNUAL CAPITAL COST ($1000)	OPERATING COST ($1000)	INSURANCE & PROPERTY TAX ($1000)	WOOD FUEL COST ($1000)	TOTAL ANNUAL COST ($1000)	ANNUAL CAPITAL COST ($1000)	OPERATING COST ($1000)	INSURANCE & PROPERTY TAX ($1000)	FOSSIL FUEL COST ($1000)	TOTAL ANNUAL COST ($1000)	ANNUAL SAVINGS ($1000)	PRESENT WORTH OF SAVINGS ($1000)
1	29.721	40.552	4.500	73.136	147.909	0.000	20.744	4.000	201.783	226.527	78.618	70.195
2	29.721	43.391	4.770	79.718	157.600	0.000	22.196	4.240	236.086	262.522	104.922	83.643
3	29.721	46.428	5.056	86.893	168.098	0.000	23.750	4.494	276.221	304.465	136.367	97.063
4	29.721	49.678	5.360	94.713	179.472	0.000	25.412	4.764	323.178	353.355	173.883	110.506
5	29.721	53.155	5.681	103.237	191.795	0.000	27.191	5.050	378.119	410.360	218.565	124.019
6	29.721	56.876	6.022	112.529	205.148	0.000	29.095	5.353	442.399	476.846	271.698	137.651
7	29.721	60.858	6.383	122.656	219.618	0.000	31.131	5.674	517.607	554.412	334.793	151.444
8	29.721	65.118	6.766	133.695	235.300	0.000	33.310	6.015	605.600	644.924	409.624	165.440
9	29.721	69.676	7.172	145.728	252.297	0.000	35.642	6.375	708.552	750.569	498.272	179.682
10	29.721	74.553	7.603	158.844	270.720	0.000	38.137	6.758	829.005	873.900	603.180	194.208
11	29.721	79.772	8.059	173.140	290.691	0.000	40.807	7.163	911.906	959.876	669.185	192.375
12	29.721	85.356	8.542	188.722	312.341	0.000	43.663	7.593	1003.096	1054.353	742.011	190.456
13	29.721	91.331	9.055	205.707	335.814	0.000	46.719	8.049	1103.406	1158.174	822.361	188.464
14	29.721	97.724	9.598	224.221	361.264	0.000	49.990	8.532	1213.747	1272.268	911.004	186.410
15	29.721	104.565	10.174	244.401	388.860	0.000	53.489	9.044	1335.121	1397.654	1008.794	184.303
16	29.721	111.884	10.785	266.397	418.786	0.000	57.233	9.586	1468.634	1535.453	1116.667	182.153
17	29.721	119.716	11.432	290.372	451.241	0.000	61.240	10.161	1615.497	1686.898	1235.657	179.966
18	29.721	128.096	12.117	316.506	486.441	0.000	65.526	10.771	1777.047	1853.344	1366.904	177.752
19	29.721	137.063	12.845	344.991	524.620	0.000	70.113	11.417	1954.751	2036.282	1511.662	175.514
20	29.721	146.657	13.615	376.041	566.034	0.000	75.021	12.102	2150.226	2237.350	1671.316	173.260

3144.502

terials handling stage was solved by plant personnel. To prevent over-size fuel (mill residue) from entering the auger feed, a shaker screen, designed in-house and constructed by plant mechanics, was placed on the front end of the fuel handling train. Sized to allow 2 in. minus fuel to pass through, this addition eliminated fuel handling problems. Accumulated fly ash and residues are cleaned from the HRT boiler on weekends when the boiler is shut down.

An unforeseen benefit of the new boiler installation (most probably attributable to the replacement with one wood-fueled unit of two fossil-fuel boilers operating in parallel) is a reduction in duration of approximately 10% in the two plant processes requiring steam.

A visual opacity reading estimated stack opacity at about 8%—well within the 20% ceiling established by the Department of Natural Resources.

Finally, the installation has performed as dependably as the fossil-fuel system it superceded.

Appendix 1
Equipment Manufacturers/Vendors

1-1. Fuel Preparation, Handling, Storage, Transport, Related Equipment
1-2. Combustion and Heat Recovery
1-3. Wood Fired Pyrolysis and Gasification Systems Equipment
1-4. Electric Power Generation Equipment
1-5. Pollution Control Equipment

1.1 FUEL PREPARATION, HANDLING, STORAGE, TRANSPORT, AND RELATED EQUIPMENT SUPPLIERS

COMPANY	PRODUCTS/SERVICES
1. Aeroglide Corporation 7100 Hillsborough Road Raleigh, NC 27602 919/851-2000	Dryers
2. Agnew Environmental Products Co. P.O. Box 1168 Grants Pass, OR 97526 503/479-3396	Briquettor for densification of dry fibrous waste into fuel or fireplace logs
3. Air-O-Flex Equipment Co. 3030 E. Hennepin Avenue Minneapolis, MN 55413 612/331-4925	Truck dumps
4. Air-Tech Industries, Inc. 85 Madison Circle Drive E. Rutherford, NJ 07073 201/460-9730	Inflatable storage equipment-air bags

NOTE: These lists are for informational purposes only. No endorsement of products and services is meant—either express or implied.

1.1 FUEL PREPARATION, HANDLING, STORAGE, TRANSPORT, AND RELATED EQUIPMENT SUPPLIERS (*Continued*)

COMPANY	PRODUCTS/SERVICES
5. American Hoist & Derrick Co. 63 S. Robert St. St. Paul, MN 55107 612/228-4321	Wood baling machinery
6. American Sheet Metal, Inc. P.O. Box 9 Tualatin, OR 97062 503/638-9611	Wood wastes, storage, and conveying systems
7. Archer Blower, Inc. 6200 SW Virginia Avenue Portland, OR 97201 503/246-7755	Wood wastes, storage, and conveying systems
8. Atlas Systems Corporation P.O. Box 11496 Spokane, WA 99211 509/535-7775	Shredded wood residue storage silos, automatic discharge systems
9. Bahco Systems, Inc. P.O. Box 48116 Atlanta, GA 30362 404/427-9051	Bark classifiers and driers, dust collectors
10. Bio-Solar Research & Dev. Corp. P.O. Box 762 Eugene, OR 503/686-0765	Wood pellets, pellet systems
11. Black Clawson, Inc. P.O. Box 1028 Everett, WA 98206 206/258-3555	Fuel preparation components and systems
12. Bocats, Inc. P.O. Box 1021 Garden City, KS 67846 316/275-7167	Live bottom and chip trailers
13. The Bonnot Company 805 Lake Street Kent, OH 44240 216/673-5829	Densified log extruders
14. California Pellet Mill Co. 1114 E. Wabash Avenue Crawfordville, IN 47933 317/362-6000	Pelletizer equipment

COMPANY	PRODUCTS/SERVICES
15. CEA, Carter Day Company 500 73rd Avenue, NE Minneapolis, MN 55430 612/571-1000	Bulk storage, wood residue handling equipment
16. Clark's Sheet Metal, Inc. P.O. Box 2428 Eugene, OR 97402 503/343-3395	Storage handling systems for wood chips and dust
17. Consolidated Baling Machine Co. 155D 7th Street Brooklyn, NY 11215 212/625-0929	Baling presses for wood residue
18. Cornell Manufacturing, Inc. Laceyville, PA 18623 717/869-1227	Wood residue handling equipment, wood splitters, hydraulic slab saws
19. Diversified Fuels 975 Oak Street Eugene, OR 97401 503/484-0371	Pelletized fuel from wood residue and equipment for making wood fuel
20. Dynamic Industries, Inc. P.O. Box 466 Barnesville, MN 56514 218/354-2211	Front end loaders
21. Ederer Inc. P.O. Box 24708 Seattle, WA 98124 206/622-4421	Rake cranes, conveyors
22. Elektroniikkayhtyma Oy SF-03100 Nummela, Finland *U.S. Distributor:* Totem Equipment Co. P.O. Box 3706 Seattle, WA 98214 206/762-9191	Metal detectors
23. Eriez Magnetics Erie, PA 16512 814/833-9881	Metal separators
24. Ferro-Tech 467 Eureka Road Wyandotte, MI 48192 313-282-7300	Briquetting equipment

1.1 FUEL PREPARATION, HANDLING, STORAGE, TRANSPORT, AND RELATED EQUIPMENT SUPPLIERS (*Continued*)

COMPANY	PRODUCTS/SERVICES
25. FMC Corporation MHS Division 3400 Walnut Street Colmar, PA 18915 215/822-0581	Fuel handling systems
26. L.B. Foster Co. P.O. Box 453 Carnegie, PA 15106 412/787-5500	Root extractors
27. Fulghum Industries, Inc. P.O. Drawer G Wadley, GA 30477 912/252-5223	Tree shears Sawmill manufacturer
28. Gebr. Weiss KG c/o Energy Control Engineering Corp. P.O. Box 3064 Charlotte, NC 704/375-1701	Wood-fired boilers
29. Goodman Equipment Corporation 4834 South Halsted Street Chicago, IL 60609 312/927-7420	Double anvil wood hog and chip mills
30. Gruendler Crusher & Pulverizer Co. 2915 N. Market Street St. Louis, MO 63106 314/531-1220	Grinders, crushers, and shredders Rock crushing manufacturer
31. Guaranty Fuels, Inc. P.O. Box 748 Independence, KS 67301 316/331-0027	Wood fuel pellets
32. Guaranty Performance Co., Inc. P.O. Box 748 Independence, KS 67301 316/331-0027	Rotary dryers, related fuel handling equipment
33. Hallco Mfg. Co., Inc. 1001½ Main Street Vancouver, WA 98660 206/696-1170	Live bottom trailers
34. Harvey Engineering & Mfg. Corp. Rt. 2, Box 478 Hot Springs, AR 71901 601/262-1010	Fuel storage, handling, and preparation systems Woodworking machinery

COMPANY	PRODUCTS/SERVICES
35. Heil Company 3000 W. Montana Milwaukee, WI 53201 414/647-3101	Dehydration equipment for predrying sludges of bark, wood, sawdust, etc.
36. Hobbs Adams Engineering Co., Inc. 1100 Holland Road Suffolk, VA 23434 804/539-0232	Hogs and related equipment Farm equipment manufacturing
37. S.W. Hooper Corporation 2111 Power Ferry Road Atlanta, GA 30339 404/955-4136	Unhogged fuel reclamation systems
38. Industrial Burner 24 W. Third Avenue Spokane, WA 99204 509/747-7965	Fuel preparation, handling systems
39. Jacksonville Blow Pipe Co. P.O. Box 3687 Jacksonville, FL 32206 904/355-5671	Wood/bark hogs, pneumatic blower systems
40. Jeffrey Manufacturing Division Dresser Industries, Inc. 500 East Morehead Street/Room 221 Charlotte, NC 28202 800/223-1954	Fuel handling, processing equipment, and hogs
41. John Deere Corporation Ottumwa, IA 515/684-4641	Crop residue densifiers
42. Kinergy Corporation 4827 Jennings Lane Louisville, KY 40218 502/964-5901	Vibrating screens, feeders, and conveyors; dust screens
43. Kockums Industries, Inc. P.O. Box 108 Trussville, AL 35173 205/655-3261	Total tree chippers
44. Koehring Canada Ltd. Box 490 Brantford, Ontario N3T 5P6 519/752-6571	Feller-bunchers, feller-forwarders
45. K-Tron Corporation P.O. Box 548 Glassboro, NJ 08028 609/881-6500	Metering conveyors

1.1 FUEL PREPARATION, HANDLING, STORAGE, TRANSPORT, AND RELATED EQUIPMENT SUPPLIERS (*Continued*)

COMPANY	PRODUCTS/SERVICES
46. Laidig, Inc. 1320 S. Merrifield Avenue Mishawaka, IN 46644 219/256-0204	Wood refuse handling systems
47. Lamb Incorporated 851 Beltline Highway Mobile, AL 36606 205/479-7401	Hogs, hammermills
48. Landers Machine Co. 207 E. Broadway Ft. Worth, TX 76104 817/336-5653	Pelletizing machinery
49. Lehigh Forming Co., Inc. P.O. Box 799 Easton, PA 18042 215/258-0830	Pellet systems
50. Mardee, Inc. 3129 E. Washington Avenue Madison, WI 53704 608/244-3331	Fuel preparation, handling storage systems
51. Maren Engineering Corp. 111 W. Taft Drive South Holland, IL 60473 312/333-6250	Baling press for wood shavings, sawdust, hydraulic baling
52. McBurney Corporation P.O. Box 47848 Atlanta, GA 30362 404/448-8144	Fuel preparation, handling systems
53. McConnell Industries P.O. Box 26210 Birmingham, AL 35226 205/942-3321	Fuel preparation, handling systems, industrial equipment manufacturing
54. M-E-C Company P.O. Box 330 Neodesha, KS 66757 316/325-2673	Rotary drum dryers, flash tube dryers, wood residue fuel preparation systems
55. Melroe Division Clark Equipment Co. Fargo, ND 58102 701-293-3220	Feller-bunchers

COMPANY	PRODUCTS/SERVICES
56. Modomekan, Inc. 2175 Parklake Drive, NE/Suite 300 Atlanta, GA 30345 404/934-3151	Wood residue storage, handling systems
57. Morbark Industries, Inc. P.O. Box 1000 Winn, MI 48896 517/866-2381	Fuel harvesting machinery, chip classification hardware
58. Munson Machinery Co., Inc. 210 Seward Avenue Utica, NY 13503 315/797-0090	Hogs, hammermills
59. Nicholson Manufacturing Co. 3670 E. Marginal Way, South Seattle, WA 98134 206/682-2752	Wood handling and preparation
60. Peabody Gordon-Piatt, Inc. P.O. Box 650 Winfield, KS 67156 316/221-4770	Fuel metering bins
61. Peerless Royal Division Royal Industries, Inc. P.O. Box 760 Paragould, AR 72450 501/236-7753	Fuel storage silos, live bottom trailers, and truck dumps
62. Piedmont Silo Company, Inc. South Dearing Road Covington, GA 30209 404/786-3031	Silos
63. Precision Chipper Corp. P.O. Box 360 Leeds, AL 35094 205/640-5181	Total tree chippers
64. Rader Pneumatics, Inc. P.O. Box 20128 Portland, OR 97220 503/255-5330	Pneumatic handling and conveying equipment and truck dumps
65. Rader Systems, Inc. 2400 Poplar Avenue/Suite 312 Memphis, TN 38112 901/761-3390	Pneumatic conveyors, disc screens
66. Rens Manufacturing Co. P.O. Box 337 Crosswell, OR 97426 503/895-2172	Metal detectors

1.1 FUEL PREPARATION, HANDLING, STORAGE, TRANSPORT, AND RELATED EQUIPMENT SUPPLIERS (*Continued*)

COMPANY	PRODUCTS/SERVICES
67. Rexnord, Inc. Vibrating Equipment Division 3400 Fern Valley Road Louisville, KY 40213	Vibrating conveyors to boiler feed
68. Reydco Trading P.O. Box 3545 Redding, CA 96001 916/347-5334	Extruded logs & machinery
69. Rome Industries Cedartown, GA 30125 404/748-7450	Feller-bunchers, skidders
70. Royer Foundry & Machine Co. Kingston, PA 18704 717/287-9624	Chippers, site preparation equipment
71. Salem Hammermill Company 2601 Industrial Drive Salem, VA 24153 703/389-8696	Industrial hammermills
72. Schutte Pulverizer Co., Inc. 61 Depot Street Buffalo, NY 14240 716/855-1555	Wood/bark grinders, air conveyors, screw elevating equipment, dumps, and hoists
73. Screw Conveyor Corporation 700 Hoffman Street Hammond, IN 46237 219/931-1450	Wood conveyors, truck dumps
74. SPM Group, Inc. 14 Inverness Drive, East Englewood, CO 80111 303/770-1201	Wood briquettes, briquetting machinery, design engineering
75. Sprout-Waldron Division of Koppers Co., Inc. 130 Logan Street Muncy, PA 17756	Wood residue storage structures, equipment for pelletizing, size classification, and materials handling
76. Stearns-Roger, Inc. P.O. Box 5888 Denver, CO 80217 303/758-1100	Rotary fuel dryers for hogged wood fuel
77. Steelcraft Corporation P.O. Box 12408 Memphis, TN 38112 901/452-5200	High/low pressure pneumatic material conveying systems, filter collectors, storage bins

COMPANY	PRODUCTS/SERVICES
78. Strong Manufacturing Co. 498 Eight Mile Road Remus, MI 49340 517/561-2280	Total tree chippers
79. Tennessee Woodex P.O. Box 10041 Knoxville, TN 37919 615/588-7411	Wood pellet sales
80. Thermal Processes, Inc. 507 Willow Springs Road LaGrange, IL 60525 312/747-6600	Fluidized-bed burners
81. TransArctic Air, Ltd. P.O. Box 11573 Vancouver, B.C., Canada 604/683-1123	Wood briquette sales, wood briquette systems
82. Triple S Dynamics 1031 S. Haskell Dallas, TX 75223 214/821-9143	Conveyors, sizing equipment
83. Union Heating, Inc. 7833 196 Southwest Edmonds, WA 98020 206/725-4588	Automatic fuel feeders for Dutch oven boilers, waste hydrogen burners
84. Weaver Star Silo, Inc. Route 4 Myerstown, PA 17067 717/866-5708	Silos
85. Wellons, Inc. P.O. Box 381 Sherwood, OR 97140 503/625-6131	Wood fuel storage bins, conveyors
86. Wesco Trailer Mfg. 1960 E. Main Street Woodland, CA 95695 916/662-9606	Chip vans
87. West Salem Machinery 665 Murlark Street Salem, OR 503/364-2213	Hogs, disc screens Sawmill equipment manufacturer

1.2 COMBUSTION AND HEAT RECOVERY EQUIPMENT SUPPLIERS

COMPANY	PRODUCTS/SERVICES
1. Abco Industries, Inc. P.O. Box 268 Abilene, TX 79604 915/677-2011	Waste burning boilers
2. Agnew Environmental Products Co. P.O. Box 1168 Grants Pass, OR 97256 503/479-3396	Wet wood refuse fired systems
3. American Fyr-Feeder Engineers 1265 Rand Road Des Plaines, IL 60016 312/298-0044	Waste burning stokers and boiler systems
4. Atlas Boiler & Equipment Co. W. 29 Spokane Falls Blvd. Spokane, WA 99211 503/747-6001	Combustion systems
5. The Babcock & Wilcox Company Power Generation Group Barberton, OH 44203 216/753-4511	Wood-fired boilers
6. Basic Environmental Engineering, Inc. 21W161 Hill Street Glen Ellyn, IL 60137 312/469-5340	Wood-fired boilers
7. Beech Island Steam, Inc. P.O. Box 681 Clearwater, SC 28922 803/593-5116	Wood-fired boilers
8. Bethlehem Corporation 25th and Lennox Street Easton, PA 18042 215/258-7111	Licensed distributor for Parkinson-Cowan multi-fuel boilers of England
9. Bigelow Company P.O. Box 706 New Haven, CT 06503 203/772-3150	Wood-fired boilers in fire tube and water tube designs
10. Biotherm Energy Systems, Inc. P.O. Box 1010 Winn, MI 48896 517/866-2381	Systems for converting wood to electric power

COMPANY	PRODUCTS/SERVICES
11. Bumstead-Woolford Co. P.O. Box 448 Woodinville, WA 98072 206/481-1600	Builders of steam systems
12. Burt Power, Inc. P.O. Box 1502 Lakeland, FL 33802 813/688-7151	Wood residue fired boilers
13. Can-Am Sales Corporation P.O. Box 158 Skaneateles, NY 13152 315/685-5611	Fluidized-bed carbonizer and gasification systems
14. Coen Company, Inc. 1510 Rollins Road Burlingame, CA 94010 415/697-0440	Wood-fired suspension and cyclone burners
15. Combustion Engineering, Inc. 1000 Prospect Hill Road Windsor, CT 06095 203/688-1911	Industrial boilers
16. Combustion Power Company, Inc. 3146 Willow Road Menlo Park, CA 94025 415/324-4744	Wood-fired fluidized beds
17. Combustion Service & Equip. Co. 2016 Babcock Blvd. Pittsburgh, PA 15209 412/821-8900	Wood-fired boilers
18. Deltak Corporation P.O. Box 9496 Minneapolis, MN 55440 612/544-3371	Wood-fired boilers
19. Detroit Stoker Company 1510 E. First Street Monroe, MI 48161 313/241-9500	Wood-fired boilers
20. Dorr-Oliver-Long 174 West Street, South Orillia, Ontario, Canada 705/325-6181	Wood-fired fluidized beds
21. Energex Ltd. P.O. Box 4208 Portland, OR 97208 503/286-8231	Wood-fired suspension and cyclone burners

1.2 COMBUSTION AND HEAT RECOVERY EQUIPMENT
SUPPLIERS (*Continued*)

COMPANY	PRODUCTS/SERVICES
22. Energy Control & Engineering 2150 Hawkins Street Charlotte, NC 28203 704/375-1701	Industrial boilers
23. Energy Products of Idaho P.O. Box 153 Coeur d'Alene, ID 83814 208/667-6439	Wood residue combustion and energy recovery systems, and fluid bed
24. Energy Resources Company 185 Alewife Brook Parkway Cambridge, MA 02138 617/661-3111	Fluidized-bed reactors
25. Energy Systems, Inc. P.O. Box 609 Hixson, TN 37343 615/870-3686	Multi-fuel boilers
26. Enertherm, Inc. P.O. Box 910 Mount Vernon, WA 98273 206/336-3167	Waste material incineration systems
27. Envirometrix, Inc. 4725 University Way, NE Seattle, WA 98115 206/524-6350	Wood residue boiler systems
28. Far Western Equip. & Chemical Co. P.O. Box 11211 Portland, OR 97211 503/287-4712	Sander dust burner systems, wood stoker systems
29. Foster Wheeler Corporation P.O. Box 807 Bellevue, WA 98009 206/454-7955	Wood residue fired boilers
30. G&S Mill 75 Otis Street Northborough, MA 01532 617/393-9266	Wood residue fired boilers
31. The Gaskell Company, Inc. P.O. Box 13255 Memphis, TN 38113 901/775-3222	Builders of steam systems

COMPANY	PRODUCTS/SERVICES
32. Gebr. Weiss K.G. % Energy Control Engineering P.O. Box 3064 Charlotte, NC 28203 704/375-1701	Wood-fired boilers
33. Gotaverden Energy Systems Ltd. 111 Railside Road/Suite 300 Toronto, Ontario, Canada M3A 1B2 416/441-2421	Wood-fired boilers
34. Guaranty Performance Co., Inc. P.O. Box 748 Independence, KS 67301 316/331-0020	Suspension and cyclone wood burners
35. A.F. Holman Boiler Works, Inc. 1956 Singleton Blvd. Dallas, TX 75212 214/637-0020	Wood residue fired boilers
36. Incinergy Systems Ltd. 402 West Pender Street Vancouver, B.C., Canada 604/684-6013	Wood-fired fluidized beds
37. Industrial boiler Company 221 Law Street Thomasville, GA 31792 912/226-3024	Wood residue burner and boiler systems
38. Industrial Burner 24 W. Third Avenue Spokane, WA 99204 509/747-7965	Wood residue burner and boiler systems
39. Johnston Boiler Company Ferrysburg, MI 49409 616/842-5050	Wood-fired fluidized beds
40. Jimaroo Ltd. 1028 Royal York Road Toronto, Ontario, Canada	Packaged steam or hot water boilers utilizing sawdust or hogged wood bark
41. E. Keeler Company 238 West Street Williamsport, PA 17701 814/326-3361	Smaller sizes wood residue boilers
42. Kewanee Boiler Corporation 101 N. Franklin Street Kewanee, IL 61442 309/853-3541	Wood residue fired boilers

1.2 COMBUSTION AND HEAT RECOVERY EQUIPMENT
SUPPLIERS (*Continued*)

COMPANY	PRODUCTS/SERVICES
43. Kipper & Sons Engineers, Inc. 2616 Western Avenue Seattle, WA 98121 206/622-4545	Wood residue fired boilers
44. Konus-Systems, Inc. 2690 Cumberland Parkway, NW Atlanta, GA 30339 404/435-2126	Wood residue fired hot oil systems
45. Lamb-Cargate Industries, Inc. 1135 Queens Avenue New Westminster, B.C., Canada 604/521-8821	Complete wood energy combustion systems
46. Lasker Boiler & Engineering Corp. 3201 S. Wolcott Avenue Chicago, IL 60608 312/523-3700	Multiple fuel-fired field-erected boiler systems
47. Lockhead Haggerty Eng. & Mfg. Co. 3904 Grant Street Burnaby, B.C., Canada 604/299-6265	Wood residue burners
48. Mardee, Inc. 3129 E. Washington Avenue Madison, WL 53704 608/244-3331	Wood residue fired boilers
49. The McBurney Corporation P.O. Box 47858 Atlanta GA 30362 404/448-8144	Builders of steam systems
50. McConnell Industries, Inc. P.O. Box 26210 Birmingham, AL 35226 205/942-3321	Wood-fired suspension and cyclone burners
51. Mechanical Equipment Company 7212 Woodlawn Avenue, NE Seattle, WA 98115 206/523-8526	Stoker-fired boilers and/or incinerators
52. Moore-Oregon Canada P.O. Box 4208 Portland, OR 97208 503/286-8231	Energex suspension burners

COMPANY	PRODUCTS/SERVICES
53. Olivine Corporation 1015 Hilton Avenue Bellingham, WA 98225 206/733-3332	Wood combustion systems
54. Peabody Gordon-Piatt, Inc. P.O. Box 650 Winfield, KS 67156 316/221-4770	Wood-fired suspension and cyclone burners
55. The Perry Smith Co., Inc. P.O. Box 8711 Chattanooga, TN 37411 615/892-7130	Builders of steam systems
56. Ray Burner Company 1301 San Jose Avenue San Francisco, CA 94112 415/333-5800	Packaged fire tube wood residue boilers and auxiliary burners
57. Rettew Automation, Inc. Box 65 Newmanstown, PA 17073 215/589-2024	Wood-fired burners
58. Riley Stoker Corporation P.O. Box 547 Worcester, MA 01613 617/852-7100	Wood-fired boilers
59. John Rogers Company 4605 Illinois Avenue Louisville, KY 41111 502/458-5400	Wood residue fired systems stokers
60. W.C. Rouse Company P.O. Box 10272 Greensboro, NC 27404 919/299-3035	Wood combustion systems
61. Steelcraft-Vyncke Corporation P.O. Box 12408 Memphis, TN 38112 901/452-8094	Wood residue fired combustion systems
62. Sunbeam Equipment Corporation 180 Mercer Street Meadville, PA 16335 814/724-1400	Wood residue fired system
63. Thermal Processes, Inc. 507 Willow Springs Road LaGrange, IL 60525 312/747-6600	Wood-fired fluidized beds

1.2 COMBUSTION AND HEAT RECOVERY EQUIPMENT SUPPLIERS (*Continued*)

COMPANY	PRODUCTS/SERVICES
64. Vastine Engineering Company Box 39, RRE Brevard, NC 287 704/884-2455	Wood residue fired burner
65. Volcano Ltd. 2020 St. Anne Street Hyacinthe, Quebec, Canada 514/774-5326	Residue burning equipment for energy conversion
66. Waycot Systems Ltd. 2940 Maine Street Vancouver, B.C., Canada 604/876-6511	Wood residue fired burners
67. Wellons, Inc. P.O. Box 381 Sherwood, OR 97140 503/625-6131	Wood residue fired steam generating plant, drying systems, boilers
68. Woodamation, Inc. P.O. Box 1365 Chalmette, LA 70044 504/279-1010	Direct fired waste burning systems
69. Wyatt Engineers, Inc. 3214 16th Avenue, SW Seattle, WA 98115 206/682-2501	Wood residue fired boilers
70. York-Shipley, Inc. P.O. Box 349 York, PA 17405 717/843-8831	Complete wood residue fired combustion systems or conversion units for existing boilers fluid bed (EPI)
71. Zurn Industries, Inc. Erie City Energy Division Erie, PA 16503 814/452-6421	Wood residue fired boilers

1.3 WOOD-FIRED PYROLYSIS AND GASIFICATION SYSTEMS EQUIPMENT SUPPLIERS

COMPANY	PRODUCTS/SERVICES
1. Alberta Industrial Developments, Ltd. 704 Cambridge Building Edmonton, Alberta, Canada T5J 1R9 403/429-4094	Gasifiers
2. Applied Engineering 1525 Charleston Highway Orangeburg, SC 29115 803/534-2424	Pyrolysis and gasification systems
3. The Biomass Corporation 951 Live Oak Blvd. Yuba City, CA 95901 916/674-7230	Pyrolysis and gasification systems
4. B.C. Research 3650 Wesbrook Mall Vancouver, B.C., Canada V6S 2L2 604/224-4331	Pyrolysis and gasification systems
5. D.M. International P.O. Box 36444 Houston, TX 77036 713/782-3440	Pyrolysis and gasification systems
6. DeKalb AgResearch, Inc. DeKalb, IL 60115 815/758-3461	Pyrolysis and gasification systems
7. Enerco, Inc. Box 139A RD#1 Langhorne, PA 19407 215/493-6565	Pyrolysis and gasification systems
8. ERCO Company, Inc. 185 Alewife Brook Parkway Cambridge, MA 02138 617/661-3111	Pyrolysis and gasification systems
9. Forest Fuels, Inc. 7 Main Street Keene, NH 03431 603/352-2865	Pyrolysis and gasification systems
10. Halcyon Associates, Inc. The Halcyon, Maple Street East Andover, NH 03231 603/735-5356	Pyrolysis and gasification systems

1.3 WOOD-FIRED PYROLYSIS AND GASIFICATION SYSTEMS EQUIPMENT SUPPLIERS (*Continued*)

COMPANY	PRODUCTS/SERVICES
11. Industrial Combustion, Inc. 4465 N. Oakland Avenue Milwaukee, WI 53211 414/332-4100	Pyrolysis and gasification systems
12. Mr. E.R. Mellenger 108 Carwarthen Street Saint John, New Brunswick, Canada 506/657-2764	Pyrolysis and gasification systems
13. Pioneer Hi-Bred International Inc. 1000 West Jefferson Street Tipton, IN 46072 317/675-4587	Pyrolysis and gasification systems
14. University of California Department of Agricultural Eng. Davis, CA 95616 916/752-1421	Pyrolysis and gasification systems
15. Westwood Polygas Ltd. (INTEG) 1155 W. Pender Street Vancouver, B.C., Canada U6E 2P4 604/682-7841	Pyrolysis and gasification systems

1.4 ELECTRIC POWER GENERATION EQUIPMENT SUPPLIERS

COMPANY	PRODUCTS/SERVICES
1. Brown & Morrison P.O. Box 240827 Charlotte, NC 28223 704/554-8470	Turbines and Systems
2. Brown Boveri Corp. 1462 Livingston Avenue North Brunswick, NJ 08902 201/932-6013	Turbines and Systems
3. Coppus Engineering Corp. 344 Park Avenue Worcester, MA 01610 617/756-8391	Turbines and Systems
4. DeLaval Turbine Inc. 852 Nottingham Way Trenton, NJ 08638 609/890-5000	Turbines and Systems
5. The Elliott Company Jeanette, PA 15644 422/527-2811	Turbines and Systems
6. General Electric Co. 1 River Road Schenectady, NY 12345 518/385-2211	Turbines and Systems
7. Terry Steam Turbine Co. Lamberton Road Windsor, CT 203/688-6211	Turbines and Systems
8. Thermo Electron Corp. Energy Systems 101 First Avenue/Dept. T Waltham, MA 02154 617/890-8700	Turbines and Systems
9. Turbodyne Corporation Steam Turbine Div./Dept. 175 Wellsville, NY 14895 716/593-1234	Turbines and Systems
10. Westinghouse Canada P.O. Box 510 Hamilton, Ontario, Canada 416/528-8811	Turbines and Systems
11. Worthington Turbine International 2965 Flowers Road, South Chamblee, GA 30341 404/451-0396	Turbines and Systems

1.5 POLLUTION CONTROL EQUIPMENT SUPPLIERS

COMPANY	PRODUCTS/SERVICES
1. Aget Manufacturing Co. P.O. Box 248 Adrian, MI 49221 517/263-5781	Mechanical collectors
2. American Air Filter Co., Inc. 215 Central Avenue Louisville, KY 40277 502/637-0011	Baghouses
3. Anacon Instruments Ltd. 1723A Bayview Avenue Toronto, Ontario, Canada 416/482-7671	Instruments for measuring fuel moisture and stack gas composition
4. Anderson 2000 Inc. P.O. Box 20769 Atlanta, GA 30320 404/997-2000	Venturi scrubbers
5. Automated Combustion P.O. Drawer 9 Lake Oswego, OR 97034 503/636-4569	Automated damper control system
6. Carothers Company P.O. Box 2748 Eugene, OR 97402 503/342-6031	Dust filtration equipment
7. CEA Carter Day Company 500 73rd Avenue, NE Minneapolis, MN 55430 612/571-1000	Air pollution equipment for wood boilers
8. Clarke's Sheet Metal, Inc. Box 3428 Eugene, OR 97402 503/343-3395	Baghouses
9. Combustion Power Company, Inc. 1346 Willow Road Menlo Park, CA 94025 415/324-4744	Dry collectors
10. Donaldson Company, Inc. P.O. Box 3217 St. Paul, MN 55165 612/698-0931	Dust collection systems
11. Enviro-Systems & Research, Inc. 2141 Patterson Avenue, SW Roanoke, VA 24016 800/631-8125	Dust collectors, scrubbers, and components

COMPANY	PRODUCTS/SERVICES
12. Far Western Equip. & Chemical Co. P.O. Box 11211 Portland, OR 97211 503/287-4712	Instrumentation and controls for burners and stokers
13. Fisher-Klosterman, Inc. P.O. Box 11190/Station H Louisville, KY 40211 502/776-1505	Wet scrubbers
14. FMC Corp. Environmental Equipment Division 1800 FMC Drive West Itasca, IL 60143 312/861-6000	Dust collectors, scrubbers and components
15. Hindson Process Equipment 1372 Main Street, N Vancouver, B.C., Canada 604/986-1251	Flow measuring devices and dewatering screens
16. Industrial Boiler Co., Inc. P.O. Box 936 Thomasville, GA 31792 912/226-3024	Pollution equipment, autofeed, controls
17. Industrial Process Equipment, Inc. P.O. Box 153 Lynnfield, MA 01940 617/245-0282	Dust control, pneumatic conveying, particle sizing equipment
18. Lear Siegler Environmental Technology Div. 74 Inverness Drive East Englewood, CO 80110 303/770-3300	Dust collectors, controls for burner monitoring devices and scrubbers
19. Maschinefabrik A. Lambion D-3548 Arolsen-Wetterburg West Germany Phone: 05961-611	Dust extractors, induced draft fans, electric control and regulating systems
20. McCarthy Products Company P.O. Box 15315 Seattle, WA 98115 206/522-1700	Continuous single and multichannel moisture detectors and controls
21. Mechanical Equipment Company 7212 Woodlawn Avenue, NE Seattle, WA 98115 206/523-8526	Combustion control systems
22. Mikro Pul Corp. 10 Chatham Road Summit, NJ 07901 201/273-6360	Baghouses

1.5 POLLUTION CONTROL EQUIPMENT SUPPLIERS (*Continued*)

COMPANY	PRODUCTS/SERVICES
23. Moisture Systems Corporation P.O. Box 97 Hopkinton, MA 01746 617/435-6881	Continuous process moisture analyzer
24. Pacific Technology, Inc. 235 Airport Way Renton, WA 98055 206/623-9080	Electrical power demand control systems
25. Peabody Gordon-Piatt, Inc. P.O. Box 650 Winfield, KS 67156 316/221-4770	Air handling systems and wet scrubbers
26. Phelps Fan Mfg. Co., Inc. P.O. Box 9588 Little Rock, AR 72219 501/565-3413	Fans and blowers
27. Redco Controls 1050 W. Ewing Street Seattle, WA 98119 206/775-1561	Control equipment for energy conversion systems
28. Research-Cottrell P.O. Box 1500 Somerville, NJ 201/685-4000	Electrostatic precipitators
29. Torit Division Donaldson Co., Inc. P.O. Box 3217 St. Paul, MN 55165 612/698-0391	Cyclones, baghouses
30. Western Precipitation Division Joy Manufacturing Co. P.O. Box 2744, Terminal Annex Los Angeles, CA 213/240-2300	Precipitators, mechanical collectors
31. Zurn Industries, Inc. Air Systems Division P.O. Box 2206 Birmingham, AL 35201 205/252-2181	Baghouses and monitoring devices

Appendix 2
STATE FORESTRY COMMISSIONS USDA FOREST SERVICE OFFICES (BY STATE)

FORESTRY

ALABAMA
Moody, Cecil W.
State Forester
Forestry Commission
513 Madison Ave.
Montgomery, AL 36104
(202) 832-6587

ALASKA
Smith, Theodore G.
Director
Div. of Forest, Land & Water
  Management
Dept. of Natural Resources
323 E. Fourth Ave.
Anchorage, AK 99501
(907) 279-5577

ARIZONA
Johnston, Kelly
Deputy State Forester
Natural Resources
  Conservation Div.
Land Dept.
1624 W. Adams
Phoenix, AZ 85007
(602) 255-4633

ARKANSAS
Gresham, Bill
Director
Forestry Commission
Dept. of Commerce
3821 W. Roosevelt Rd.
Little Rock, AR 72204
(501) 371-1733

DELAWARE
Gabel, Walter F.
State Forester
Forestry Section
Div. of Production &
  Promotion
Dept. of Agriculture
Agriculture Bldg.
Dover, DE 19901
(302) 678-4815

FLORIDA
Bethea, John M.
Director
Div. of Forestry
Dept. of Agriculture &
  Consumer Services
Collins Bldg.
Tallahassee, FL 32304
(904) 488-4274

GEORGIA
Shirley, A. Ray
Director
Forestry Commission
P.O. Box 819
Macon, GA 31202
(404) 744-3237

HAWAII
Landgraf, Libert K.
State Forester
Forestry Div.
Dept. of Land & Natural
  Resources
1151 Punchbowl St.
Honolulu, HI 96813

IOWA
Hertel, Gene
Forester
Conservation Commission
300 Fourth St.
Des Moines, LA 50319
(515) 281-5629

KANSAS
Conley, Jerry
Director
Forestry, Fish & Game
  Commission
P.O. Box 1028
Pratt, KS 67124
(316) 672-5911

KENTUCKY
Grimm, Elmore C.
Director
Div. of Forestry
Bureau of Natural Resources
Dept. for Natural Resources &
  Environmental Protection
Capital Plaza Tower
Frankfort, KY 40601
(502) 564-4496

LOUISIANA
McFatter, D. L.
Asst. Secretary/State Forester
Office of Forestry
Dept. of Natural Resources
P.O. Box 1628
Baton Rouge, LA 70821
(504) 925-4510

CALIFORNIA
Personen, David E.
Director
Dept. of Forestry
Resources Agency
Rm. 1505, Ninth St.
Sacramento, CA 95814
(916) 445-3976

COLORADO
Borden, Tom
State Forester
Forest Service
Colorado State Univ.
Fort Collins, CO 80521
(303) 491-6304

CONNECTICUT
Garrepy, Robert
State Forester
Forest Unit
Dept. of Environmental
  Protection
Rm. 260, 165 Capitol Ave.
Hartford, CT 06115
(203) 566-5348

MICHIGAN
Webster, Henry H.
Chief
Forest Management Div.
Dept. of Natural Resources
5th Fl., Mason Bldg.
P.O. Box 30028
Lansing, MI 48909
(517) 373-1275

MINNESOTA
Hitchcock, Raymond B.
Director
Div. of Forestry
Dept. of Natural Resources
658 Cedar St.
St. Paul, MN 55155
(612) 296-2894

IDAHO
Gillette, Jack E.
Asst. Director
Div. of Forest Resources
Dept. of Lands
121 State House
Boise, ID 83720
(208) 384-3282

ILLINOIS
Mickelson, Allan
State Forester
Div. of Forestry
Dept. of Conservation
RR 5, Conservation Area
Springfield, IL 62707
(217) 782-2361

INDIANA
Datena, John F.
State Forester
Forestry Div.
Dept. of Natural Resources
613 State Office Bldg.
Indianapolis, IN 46204
(317) 232-4105

NEW HAMPSHIRE
Natti, Theodore
Director
Forest & Lands Div.
Dept. of Resources &
  Economic Development
310 State House Annex
Concord, NH 03301
(603) 271-2214

NEW JERSEY
Bamford, Gordon T.
Chief
Bureau of Forestry
Div. of Parks & Forestry
Dept. of Environmental
  Protection
P.O. Box 1420
Trenton, NJ 08625
(609) 292-2520

MAINE
(Vacancy)
Director
Bureau of Forestry
Dept. of Conservation
State House
Augusta, ME 04333
(207) 289-2791

MARYLAND
MacLauchlan, Donald E.
Director
Forest Service
Dept. of Natural Resources
Tawes State Office Bldg.
Annapolis, MD 21401
(301) 269-3776

MASSACHUSETTS
Bliss, Gilbert
Director
Forest & Parks Div.
Dept. of Environmental
  Management
Executive Office of
  Environmental Affairs
100 Cambridge St.
Boston, MA 02202
(617) 727-3180

OKLAHOMA
Riley, Jim
Director
Div. of Forestry
Dept. of Agriculture
122 State Capitol
Oklahoma City, OK 73105
(405) 521-3886

OREGON
Schroeder, J.E.
Forester
Dept. of Forestry
2600 State St.
Salem, OR 97310
(503) 378-2511

MISSISSIPPI
Holman, Jack
State Forester
Forestry Commission
908 Robert E. Lee Bldg.
Jackson, MS 39205
(601) 354-7124

MISSOURI
Presley, Jerry J.
State Forester
Forestry Div.
Dept. of Conservation
2901 N. Ten Mile Dr.
Jefferson City, MO 65101
(314) 751-4115

MONTANA
Pyke, Larry
Area Supervisor
Forest Resources Div.
Dept. of Natural Resources &
  Conservation
8001 N. Montana Ave.
Helena, MT 59601
(406) 449-3633

NEBRASKA
Ferrill, Mitchell D.
Forester
Dept. of Forestry
University of Nebraska
201 Miller Hall
Lincoln, NE 68583
(402) 472-2944

NEVADA
Smith, Lowell V.
Forester
Div. of Forestry
Dept. of Conservation &
  Natural Resources
Rm. 330, Nye Bldg.
Carson City, NV 89710
(702) 885-4350

NEW MEXICO
Gallegos, Raymond R.
State Forester
Natural Resources Dept.
Land Office Bldg.
Santa Fe, NM 87501
(505) 827-2312

NEW YORK
Van Valkenburgh, Norman
Director
Div. of Lands & Forests
Dept. of Environmental
  Conservation
50 Wolf Rd.
Albany, NY 12233
(518) 457-7430

NORTH CAROLINA
Winkworth, R. C.
Director
Div. of Forest Resources
Dept. of Natural Resources &
  Community Development
512 N. Salisbury St.
Raleigh, NC 27611
(919) 733-2162

NORTH DAKOTA
Johnson, Robert
State Forester
School of Forestry
North Dakota State Univ.
Bottineau, ND 58318
(701) 228-2277

OHIO
Gebhart, Ernest J.
Chief
Div. of Forestry
Dept. of Natural Resources
Fountain Square
Columbus, OH 43224
(614) 466-7842

PENNSYLVANIA
Thorpe, Richard R.
Director
Bureau of Forestry
Dept. of Environmental
  Resources
100 Evangelical Press Bl
Harrisburg, PA 17105
(717) 787-2708

RHODE ISLAND
Deion, Henry J., Jr.
Chief
Div. of Forest Environment
Dept. of Environmental
  Management
Old Hartford Pike
Scituate, RI 02857
(401) 647-3367

SOUTH CAROLINA
Kilian, Leonard A., Jr.
State Forester
Forestry Commission
5500 Broad River Rd.
P.O. Box 21707
Columbia, SC 29211
(803) 758-2261

SOUTH DAKOTA
Verville, Jim
Director
Div. of Forestry
Dept. of Game, Fish
Anderson Bldg.
Pierre, SD 57501
(605) 773-3623

TENNESSEE
Young, Max J.
Forester
Div. of Forestry
Dept. of Conservation
4711 Trousdale Dr.
Nashville, TN 37211
(615) 741-3326

TEXAS
Kramer, Paul R.
State Forester
Forest Service
Texas A&M University
302 System Administration
  Bldg.
College Station, TX 77843
(713) 845-2641

UTAH
Walker, Edward
Director
Div. of Forests
Dept. of Forests, Parks &
  Recreation
Agency of Environmental
  Conservation
79 River St.
Montpelier, VT 05602
(802) 828-3375

WASHINGTON
Morton, L. V.
Supervisor
Div. of Forest Land
  Management
Dept. of Natural Resources
Public Lands Bldg.
Olympia, WA 98504
(206) 753-0671

WYOMING
Johnson, Carl
Supervisor
Forestry div.
Public Lands
State Capitol
Cheyenne, WY 82002
(307) 777-7331

VERMONT
Walker, Edward
Director
Div. of Forests
Dept. of Forests, Park &
  Recreation
Agency of Environmental
  Conservation
79 River St.
Montpelier, VT 05602
(802) 828-3375

WEST VIRGINIA
Kelly, A. W., Jr.
State Forester
Div. of Forestry
Dept. of Natural Resources
Rm. 732, Bldg. 3
State Capitol Complex
Charleston, WV 25305
(304) 348-2788

PUERTO RICO
Soltero-Harrington, Fred
Secretary
Dept. of Natural Resources
P.O. Box 5887
San Juan, PR 00906
(809) 723-3090

VIRGINIA
Custard, Wallace F.
State Forester
Div. of Forestry
Dept. of Conservation &
  Economic Development
P.O. Box 3758
Charlottesville, VA 2290
(804) 977-6555

WISCONSIN
Reinke, Milton E.
Director
Bureau of Forestry
Div. of Resource Management
Dept. of Natural Resources
4610 University Ave.
Madison, WI 53702
(608) 266-0842

Appendix 3
1981 Wood Fuel Supply Survey— Georgia

1981 WOOD RESIDUE SURVEY

Prior to this particular study, it was felt that there was not adequate information about the wood mill residue available in the state of Georgia. Some telephone calls had been made for particular areas, and another study existed but which was not complete. The companies in the SIC codes that were appropriate were found in the Manufacturers Directory. About 370 companies fell into the categories which were of interest. To approach these companies in a professional manner, it was decided that a well-designed form for them to fill out was necessary. The form was designed to be simple and short so as to allow the respondent to complete the form in ten minutes or less if the data were close at hand. A long form might have reduced the overall return rate. It was important to place the respondent's name on the form beforehand—reducing the time it took for him to fill it out; also, there would be no question from where the form came. The form and survey results that were developed follow.

As in any questionnaire, there were some who did not appear to understand the questions that were asked, but in this particular type of canvass, it was felt that some information and contact with the residue suppliers was necessary.

In order to enhance the response, each letter was addressed to the particular person listed in the Manufacturers Directory, with his name on the inside salutation of the letter itself. The letters were sent out with a self-addressed, stamped return envelope. The response level to this survey was 40%, which was felt to be very significant.

The survey was sent out to 368 wood products companies across the state of Georgia during January 1981. These companies were in SIC Code 2421 as determined by the Manufacturers Directory. There were 145 companies that responded.

This survey is probably very conservative. Many of the larger companies whose residue is already under contract did not respond. One of the major problems for a sawmill operator is determining how much residue he has, especially if he is not currently selling waste. This survey may only be addressing 30% to 40% of the waste that is available.

Considering the downturn of the construction business, this wood mill residue supply picture should improve as the economy recovers.

In the survey, the companies were asked if they wished to be listed as a supplier in our publications. If they did not wish to be listed, we took their name and amount off the company list and only included them in the county total. This means that the company listing amount and the company total may not always be the same.

To interpret the survey, a few items must be clarified. The column "Waste Used For Fuel" asks whether the company is using its own waste for fuel at its own plant. If the company said "Y" or *yes,* then it would specify "B" (*boiler*) or "D" (*dryer*).

The "waste to sell" column asks if the company currently has available waste to sell, i.e., "Y" (*yes*) or "N" (*no*).

"Selling waste" asks if the company is currently selling waste and in what form it is selling it. In this case, "WT" stands for *whole tree chips.* "BK" stands for *bark* and "S" stands for *sawdust* or *shavings.*

In the quantity column, "NQG" stands for *no quantity given.* The numbers are given in tons/year. These are divided up into *chips* and *other.* The "other" is bark, sawdust, shavings, or slabs.

The average delivered price across the state for "other" is $9/ton and the average price for delivered chips is $17.50/ton.

SPECIMEN

Woody Birch Pallet Co.
Timber Springs
Georgia 11111
ATTN: Mr. Woody Birch

Years in business at this location ________________ . Number of employees ____________
Responding Person ________________________________Title ________________________________
Nature of Business (Lumber Co., Sawmill, Veneer, Pallet, etc.)
____

Volume of Business (tons per day, board feet of lumber, etc.)
____

Is Wood Waste Used for Fuel Here? Yes __________ No __________
If used, how?
   Boiler: Yes __________
   Electrical Generation: Yes __________
   Dryer: Yes __________
   Other (Describe) ____
____

Is wood waste available to sell? Yes __________ No __________
Are you currently selling wood waste? Yes __________ No __________
What % of production is available for sale currently? __________
In what form is it available:

	How much (tons/day)	Price/ton (if sold)
Bark	__________	__________
Sawdust	__________	__________
Chips	__________	__________
Shavings	__________	__________

Is the price given delivered? Yes __________ No __________
Do you want to be identified as a specific source of mill residue in our publications?
Yes __________ No __________

May Georgia Tech visit this location? Yes _____________ No _____________
Whom should we contact for a plant visit? ____
Your cooperation is appreciated.
Please put in the self-addressed stamped envelope and return.

Name	Waste Used for fuel	Waste to Sell	Selling Waste	Quantity Tons/Yr. Other	Chips
APPLING COUNTY					
Baxley Veener and Cleat Co.					
Dixon's Forest Enterprises Inc.					
Knight Lumber Co.	N	Y	Y	NQG	NQG
Youmans Timber Co., Inc.					
ATKINSON COUNTY					
Pearson Wood Division	Y-B	N	Y		
Googe and Bridges Lumber Co.					
McCranie Bros. Wood Pres., Inc.	Y-B-D	N	N		
BALDWIN COUNTY					
T and S Hardwood					
Purcell Bros.	N	Y	Y-WT		28,000
Dixiewood Inc.	N				
BANKS COUNTY					
Padco Manufacturing Co.					
Garrison and Sons					
Skitts Mountain Lumber Co.					
Appalachian Industries					
BARROW COUNTY					
Hardegree Lumber Co.					
BARTOW COUNTY					
Knight Mercantile Co.					
BEN HILL					
Empire Forest Products	N	Y	Y-Bk	14,000	
Forest Prod. Div.					
Gilman Paper					
ITT Rayonier, Inc.					
Meredith Pole and Timber					
Tolleson Lumber					
BERRIEN COUNTY					
Moore Humbert Lumber					
Nashville Variety Works					
BIBB COUNTY					
True Temper Corp.					
Southern Crate and Veneer					
Dependable Lumber Co. in Macon					
Emerick Cabinets					
Franline Cabinet and Millwork					

Name	Waste Used for fuel	Waste to Sell	Selling Waste	Quantity Tons/Yr. Other	Chips
Georgia Timberlands	N	Y	Y-WT		21,000
Huttig Sash and Door					
Marshall Lumber					
Southern Wood Piedmont Co.					
Willingham Sash and Door	N	N	N		
BLECKLEY COUNTY					
Faulk Lumber					
Griffin-Porter Lumber Co.					
Bleckley Lumber Co.	Y-B	Y	Y	2,240	5,600
BROOKS COUNTY					
H and G Lumber Co.	N	Y	Y	2,800	4,200
W.B. Jones and Son					
BRYAN COUNTY					
Centennial Lumber Co.	N	Y	Y	22,400	168,000
BULLOCH COUNTY					
W.K. Jones Lumber					
Kendrick Woodyard, Inc.					
W.M. Sheppard Lumber					
Continental Forest Inc.					
Claude Howard Lumber Co.	Y-D	Y	Y	25,200	39,200
BURKE COUNTY					
Sardis Lumber	N	Y	Y	3,360	5,040
BUTTS COUNTY					
W.A. Bunch, Son Lumber					
Towaliga Wood					
CAMDEN COUNTY					
Pallet Services					
Exco Company	N	Y	Y-B	1,740	
St. Regis Allied Operations					
CANDLER COUNTY					
Metter Lumber Co.	N	Y	N-S	2,800	
CARROLL COUNTY					
Richards Lumber Co.	Y-D	N	N		
Temple Manufacturing					
Precision Panel Prod. Inc.					
Villa Rica Lumber					
Southwire Wood Products					
CATOOSA					
Babb Lumber					
CHARLTON COUNTY					
Thomas Lumber and Timber					
Toledo Manufacturing Co.					

Name	Waste Used for fuel	Waste to Sell	Selling Waste	Quantity Tons/Yr. Other	Chips
CHATHAM COUNTY					
Shearouse Lumber					
Bradley Plywood					
Buie's Millwork					
Continental Forest Inc.	N	N	N		
Guerry Lumber Co.	N	Y	Y	980	420
Highsmith Lumber Co.					
Hughes Ball Co.	N	Y	N	NQG	
McCullough Enterprise					
Neal-Blun Co.					
XYLO Inc.	N	Y	Y	560	
Woodcraft Cabinet					
CHATOOGA COUNTY					
Menlo Lumber Co.					
J.P. Smith Lumber Co.					
Waiter Haney Lumber Co.					
CHEROKEE COUNTY					
J.P. Haynes Lumber Co.					
Brackin Tie, Lumber and Chip					
McKinney Wood Prop.					
CLARKE COUNTY					
Alexander Wood Products	Y-B	Y	N-Bk	2,800	
Armstrong and Dobbs	N	Y	N	28	
Athens Lumber Co.					
Hogan Lumber Co.	N	N	N		
Tanner Lumber Co.	Y	N	N		
CLAY COUNTY					
Phillips Lumber Co.					
CLAYTON COUNTY					
Clayton Hardwoods	N	Y	Y	840	
Conley Millwork and Fixture Co.					
Southern Door and Plywood					
ECCO Inc.					
Southwest Forest					
CLINCH COUNTY					
Union Timber					
St. Regis Paper					
COBB COUNTY					
Kennesaw Lumber Co.					
Acme Lumber Co.					
North American Container	N	Y	N	840	
Personality Home Manufacturers	N	Y	N	NQG	
W.P. Stephens Lumber Co.					

Name	Waste Used for fuel	Waste to Sell	Selling Waste	Quantity Tons/Yr. Other	Chips
Trimline Sales					
Howard Lumber and Kiln					
COFFEE COUNTY					
Stubbs and Southern Lumber Co.					
T and T Lumber Co.					
Union Camp	N	N	N		
Comet Enterprises	N	Y	Y	28	
FAB TEC Inc.	Y-B-D	Y	Y	21,000	36,400
COLQUITT COUNTY					
Beadles Lumber Co.	N	Y	Y	23,240	40,600
A and S Millworks					
Georgia Forest Products					
Moultrie Wood Preserving	Y	N	N		
COLUMBIA COUNTY					
COOK COUNTY					
Del-Cook Lumber Co.					
Pioneer Processors					
Triple S Pallet					
Weyerhauser					
COWETA COUNTY					
Hardwoods of Georgia	N	Y	Y	11,200	28,000
Alsobrook Lumber Co.					
CRAWFORD COUNTY					
Spillers Bros. Lumber Co.	N	Y	Y	12,600	28,000
CRISP COUNTY					
Cordele Sash Door and Lumber Co.					
DADE COUNTY					
Dave T. Brown Lumber Co.					
Clark Lumber Co.					
Dyer Lumber Co.					
DAWSON COUNTY					
Cowart Brothers					
DECATUR COUNTY					
DEKALB COUNTY					
W.E. Williams					
American Door					
Gillespie Deka Corp.					
Vorco Wood Prod.					
Product Packaging Corp.					
Tucker Door and Trim					

Name	Waste Used for fuel	Waste to Sell	Selling Waste	Quantity Tons/Yr. Other	Chips
DODGE COUNTY					
McCranie Lumber Co.					
M.C. Jones Lumber Co.					
M.C. Jones Chip Mill					
DOOLY COUNTY					
DOUGHERTY COUNTY					
Giles Builders Supply					
Smith Lumber					
Watkins Lumber	Y	Y	Y	14,000	17,360
DOUGLAS COUNTY					
Butler Pallet and Crate					
EARLY COUNTY					
Great Southern Plywood	Y-B	N	N		
W.E. McDowell Pulpwood	Y-B	N	N		
EFFINGHAM COUNTY					
J.W. Exley Lumber	N	Y	Y	8,400	
Goshen Tie and Chips	N	Y	Y	11,200	84,000
ELBERT COUNTY					
Beaver Dam Lumber					
EMMANUEL COUNTY					
Boulineau's Cabinet Shop					
St. Regis	N	Y	Y	56,000	
Yeomans and Brown Truss					
EVANS COUNTY					
O.H. Daniel Lumber Co.					
W.R. Deloach Pulpwood Co.					
Georgia Pacific					
Randall Lewis Lumber Co.	N	Y	Y	5,600	
FAYETTE COUNTY					
John Lampkins, Inc.					
FLOYD COUNTY					
Shortleaf Lumber Co.					
Jack Gresham Inc.	N	Y	Y	6,160	5,600
Smith-Evans Lumber Co.					
Cordle Bros. Lumber					
S.L. Miller and Sons					
S.I. Storey Lumber Co.					
O'Neil Mfg. Co.					
Rome Builders Supply, Inc.					
Georgia Kraft	Y-B	Y	Y-WT	18,760	

Name	Waste Used for fuel	Waste to Sell	Selling Waste	Quantity Tons/Yr.	
				Other	Chips
FORSYTH COUNTY					
Byron Anderson Sawmill					
Mr. R.W. Norwood Basket and Crate					
FRANKLIN COUNTY					
Lion Bonds Tree Co.					
Bon Tree Co.					
Harbin Homes	N	Y	Y	560	
FULTON COUNTY					
Mr. A.E. Dinsmore					
McClure Bros. Lumber	N	Y	Y	11,200	14,000
Addison-Rudesal					
Anderson McGriff					
Atlanta Hardwood Corp.					
Crown Door Corp.					
Houston Lumber Co.					
Huttig Sash and Door					
J.A. Woodcraft					
Joanna Western Mills					
Norjoe Timber					
Randall Bros.	N	N	N		
Schaeffer Sash and Door Co.					
Smith Lawrence Planning Mill					
Southern Doorlite					
West Enterprises					
Williams Bros. Lumber Co.	N	Y	Y	38,080	112,000
W.C. Meredith Co.					
Miller Lumber Co.					
B and P Lumber Co.					
Atlanta Pallet Exchange					
GILMER COUNTY					
Ellijay Lumber Co.	Y	Y	Y	1,680	2,800
Garland Lumber Co.					
Appalachian Southern Corp.					
GLASCOCK COUNTY					
Pulliam Lumber Co.	N	Y	Y	20,440	56,000
GLYNN COUNTY					
E.P. Edgy Planing Mill					
Edgy and Wooten Lumber	N	Y	Y	5,600	5,600
Lang Planning Mill					
South Georgia Millworks					
GORDON COUNTY					
Calhoun Lumber and Supply					

| | | | | Quantity Tons/Yr. | |
Name	Waste Used for fuel	Waste to Sell	Selling Waste	Other	Chips
GRADY COUNTY					
Lacey Lumber Co.	N	Y	Y	9,800	7,000
Faircloth Logging and Timber Co.					
Graco Fertilizer					
Lewis Lumber Co.	N	Y	Y	1,680	
GREENE COUNTY					
Greensboro Lumber Co					
Union Point Lumber Co.					
GWINNETT COUNTY					
Gasner Lumber Co.	N	Y	Y	8,120	7,000
Suwanee Lumber Co.					
S and S Lumber Co.					
Pallets Inc.	N	Y	N	3,360	
Aeon Box Co.					
Magbee Lumber Co.					
Charles Greer Lumber Co.	N	Y	Y	5,000	
HABERSHAM COUNTY					
Church Lumber Co.					
Wilbanks Lumber Co.					
Cornelia Veneer Co.					
Turner Lumber Co.					
Kenneth Cantrell Lumber					
Owen Cantrell Sawmill					
Mt. Airy Wood Preserving Co.					
McCollum Mfg. Co.					
Mize Lumber Co.					
Jesse L. Jones Co.					
HALL COUNTY					
Highland Millwork and Supply					
Lawson and White Lumber Co.					
HANCOCK COUNTY					
Edwards Lumber Co.					
HARALSON COUNTY					
C.N. Owens Sawmill	N	Y	Y	7,000	7,000
Pete's Cabinet					
HARRIS COUNTY					
Meriweather Lumber Co.					
Georgia Pacific					
Kraftwood Inc.					
Pineco Inc.	N	Y	Y	1,600	
HART COUNTY					
Hartwell Lumber Co.	N	Y	Y	3,640	280
Vickery Lumber Co.					

Name	Waste Used for fuel	Waste to Sell	Selling Waste	Quantity Tons/Yr. Other	Chips
HEARD COUNTY					
H and R Lumber					
HENRY COUNTY					
Shockley Forest Industries					
Metropolitan Pallet Co.	N	Y	Y	280	
Gang-Nail Structures					
HOUSTON COUNTY					
Tolleson Lumber Co.	Y-B-D	N	N		
JACKSON COUNTY					
Barnett Sawmill					
Rhodes Lumber Co.	N	Y	Y	1,680	1,680
JASPER COUNTY					
Champion Lumber Co.					
Caylor					
Georgia Pacific					
JEFF DAVIS COUNTY					
Continental Forest Inc.					
Hazelhurst Lumber and Supply					
Ridge Pallets	N	Y	Y		
Thompson Timber Co.,					
St. Regis Allied	Y	Y	N	N	
JEFFERSON COUNTY					
J.P. Morgan Construction Co.					
Battle Lumber Co.	N	Y	Y	8,400	33,600
Lamb Bros Lumber Co.	Y	Y	Y	8,400	8,400
Norton Lumber and Supply Co.					
JENKINS COUNTY					
Miller Lumber Supply Co.					
JOHNSON COUNTY					
Wrightsville Lumber Co.	N	Y	Y		
JONES COUNTY					
Amay Forest Products					
LAMAR COUNTY					
Wallmark Furniture	N	Y	N		
LAURENS COUNTY					
Dublin Hardwood	N	Y	Y	8,400	
Winter Naval Stores					
Gilman Paper Co.					
Alexander Lumber					
LIBERTY COUNTY					
Koppers Company					

Name	Waste Used for fuel	Waste to Sell	Selling Waste	Quantity Tons/Yr. Other	Chips
LINCOLN COUNTY					
Aycock Bros. Lumber	N	Y	Y	19,600	56,000
LOWNDES COUNTY					
Bray Lumber Co.					
J.T. Griffin Lumber Co.					
Lanedale Co.					
Pan American Gyro-Tex					
Pre Cuts of Georgia					
LUMPKIN COUNTY					
Dahlonega Wood Works	N	Y	Y		2,240
MADISON COUNTY					
Broad River Sawmill					
Lee Lumber Co.					
Danielsville Wood Products	N	Y	Y	560	
McDUFFIE COUNTY					
McNeill Lauff and McNeill					
Temple Eastex Inc.					
Thomson Oak Flooring	Y-B	Y	Y	28,000	11,200
MERIWETHER COUNTY					
J.O. Barber Co.					
Georgia Kraft	Y-B	Y	Y	11,200	140,000
Manchester Veneer Co.	Y-B-D	Y	Y	2,800	7,000
Great Southern Paper Co.					
Georgia Pacific					
Mr. W.H. Reeves Lumber Co.					
MITCHELL COUNTY					
Alexander-Harris Co.					
Mineo Industries					
Keadle Bros Lumber					
Escambia Treating					
Southern Wood Piedmont					
Whaley Lumber Co.					
MONROE COUNTY					
Vaughn Lumber Co	N	Y	Y		
Georgia Timberlands	N	Y	Y-WT		140,000
MONTGOMERY COUNTY					
Thompson, H.V. and T.G. Lumber					
Champion Truss					
MORGAN COUNTY					
Underwood and Son					
Oconee Hardware Lumber	N	Y	Y	10,080	19,600
Alumni Furniture					

Name	Waste Used for fuel	Waste to Sell	Selling Waste	Quantity Tons/Yr. Other	Chips
MURRAY COUNTY					
Sutton Lumber Co.					
Sluder Lumber Co.					
Fort Mountain Lumber					
MUSCOGEE COUNTY					
Columbus Cabinet					
Duncan and Johnson	N	Y	Y	8,400	3,360
Millwork Manufacturer					
Valley Components	Y	Y	N	140	
Wells Lumber Co.	N	Y	Y	560	
Woodcraft By Frank MacDonald					
Harvey Lumber					
NEWTON COUNTY					
Morgan Timber					
Williams Brothers Lumber					
OCONEE COUNTY					
Bill Wood Co.					
OGLETHORPE COUNTY					
Span-It Inc.					
PAULDING COUNTY					
Black Sawmill					
PWK Supply Co.					
Raleigh Wilson Lumber Co.	N	Y	Y		
Reese Walraven Lumber Co.					
Fred Henson Lumber Co.					
PEACH COUNTY					
Boyd's Industrial Woodwork	Y	Y	N		
Paneling Industries					
PIERCE COUNTY					
Gilman Paper					
POLK COUNTY					
Paynes Planing Mill					
Coastal Timberlands Inc.					
Cedartown Paper Board					
Moates Manufacturing Co.					
Victor Veneer					
PUTNAM COUNTY					
International Paper Co.					
Dodson Lumber Co.					
Putnam Lumber Co.					
RABUN COUNTY					
Keeners Sawmill					
Arlington Box Co., Inc.					

Name	Waste Used for fuel	Waste to Sell	Selling Waste	Quantity Tons/Yr. Other	Chips
Burgin Lumber Co.					
Day Plywood, Inc.					
Highland Timber					
Randolph County					
Day Plywood	N	N	N		
RICHMOND COUNTY					
Augusta Box and Crate Co.					
Augusta Lumber Co.					
Augusta Sash and Door					
Augusta Wood Preserving					
Coastal Lumber					
Continental Forest					
Howard Lumber Co.					
Mulherin Lumber Co.					
Bolivia Lumber					
ROCKDALE COUNTY					
J.T. Hicks Lumber Co.					
Still Lumber and Coal					
SCREVIN COUNTY					
Corbett Plywood	N	Y	Y	9,520	15,900
Evans Timber					
H and R Fries Log and Lumber					
Mobley Lumber Co.					
Pfeiffer Brothers Lumber	N	N	N		
SPAULDING COUNTY					
Orchard Hill Lumber Co.	N	N	N		
Lawrence Smith Planing Mill					
STEPHENS COUNTY					
General Doors					
STEWART COUNTY					
Greentree Manufacturing Co.	N	Y	Y	4,200	
St. Regis Allied Operation					
St. Regis Paper Co.,					
Dudley Pallet and Box Co., Inc.					
Valley Wood					
SUMTER COUNTY					
Americus Wood Preserving	N	N	N		
Shiver Lumber Co.					
Textone Inc.	N	N	N		
TALIAFERRO COUNTY					
Wisham and Hall Lumber Co.	N	N	N		
TATTNAIL COUNTY					
Earl Wilson Lumber Co.					
Ajax Forest Prod.					

Name	Waste Used for fuel	Waste to Sell	Selling Waste	Quantity Tons/Yr. Other	Chips
TELFAIR COUNTY					
Myers Hardwood Co.					
St. Regis Allied					
Georgia Pacific	N	Y	Y	11,200	
M.C. Jones Lumber					
THOMAS COUNTY					
Baker Lumber Co.					
Georgia Crate	N	Y	Y	15,960	
Georgia Wood Industries, Inc.					
Secrest Pulpwood and Timber					
TIFT COUNTY					
Riddle and Slack Lumber Co.					
Tifton Lumber Co.					
TOOMBS COUNTY					
Cato Timber					
Atlantic Wood Industries, Inc.					
Brown Thrasher, Inc.	N	Y	N		
TOWNES COUNTY					
Atkins Sawnee					
TROUP COUNTY					
J.D. Glaney	N	Y	Y	5,600	7,000
Trammel Lumber Co.	Y	Y	Y	28,000	
Trammel Lumber Co. (Chipmill)					
Daniel Lumber Co.	Y	Y	Y	NQG	
TURNER					
Weyerhaeuser Corp.					
TWIGGS COUNTY					
Knight Bros (Sawmill)					
Knight Bros (Tree Chipper)					
UNION COUNTY					
Union Lumber Co.					
Spiva Lumber Co.					
UPSON COUNTY					
Frank Buford Supply					
Keadle Lumber Enterprises	N	Y	Y	89,600	75,600
Georgia Kraft Co.					
WALKER COUNTY					
Ray Junkins Sawmill					
Gilstrap Sawmill					
WALTON COUNTY					
Atlanta Southern Corp.					
Harrison Manufacturing Co.					

Name	Waste Used for fuel	Waste to Sell	Selling Waste	Quantity Tons/Yr. Other	Chips
WARE COUNTY					
Manor Timber					
ITT Rayonier, Inc.					
St. Regis Paper—Allied Operations					
WARREN COUNTY					
Georgia Pacific					
ITT Rayonier					
WASHINGTON COUNTY					
Corbett Box Co.					
Hodges Lumber Co.					
WAYNE COUNTY					
Perkins Hardwood Lumber					
Wayne Forest Enterprises					
WEBSTER COUNTY					
Sullivan Lumber Co.					
WHITE COUNTY					
Mt. Yonah Lumber Co.					
Blalock Lumber Co.	N	Y	Y		
Freeman Lumber Co.	N	Y	Y	10,640	6,160
Irvin Lumber Co.	N	Y	Y	9,800	
Hogan Lumber Co.					
Appalachian Trail Co.					
International Hardwoods	N	Y	Y	9,800	7,000
WHITFIELD COUNTY					
Baldridge Bros.					
Deforest Lumber Co.					
Cockburn Bros. Lumber					
W.D. Cline Lumber Co.					
Mr. J.H. Langley					
WILKES COUNTY					
Continental Can					
AAA Log Homes					
Burt Lumber Co.					
Conwall					
WILKINSON COUNTY					
Culpepper Wood Products	N	Y	Y	2,800	7,000
Lord Lumber Co.					
Fountain Pallet Co.					
Bayley Veneer and Cleat Co.	N	Y	Y	11,200	21,000
Shepard Bros.					
Freeman Lumber Co.					

Appendix 4
Bibliography and References

BIBLIOGRAPHY AND REFERENCES

1. *Accident Prevention Manual for Industrial Operations,* Seventh Edition, 1978.
2. Allen, Donald E., "Fire in the Chip Pile! What to Do," *Pulp and Paper,* July 1, 1968, p. 33.
3. Allison, R.C., *Estimating Wood Fuel Production at Sawmills,* Extension Wood Products Section, School of Forest Resources, NC State University, Raleigh, NC.
4. "American National Standards Institute," 1430 Broadway, New York, NY, 10081.
5. Applefield, Milton, *The Production and Marketing of Pine Wood Residues by Small Sawmills,* Bulletin No. 50, Forest Products Department, Texas Forest Service, Lufkin, TX, 14 pp., March 1960.
6. *ASHRAE Handbook 1977-Fundamentals,* American Society of Heating, Refrigerating, and Air Conditioning Engineers, Inc., New York, NY, 1978.
7. Assarsson Anders, Croon Ingemar, Frisk Edvin, "Outside Chip Storage (OCS)" *Svensk Papperstidning,* August 1970.
8. Baumeister, Theodore, *Marks' Standard Handbook for Mechanical Engineers,* Ch. 7, McGraw-Hill Book Co., New York, NY, 1978.
9. Bergman, Osten, "Thermal Degradation and Spontaneous Ignition in Outdoor Chip Storage," *Svensk Papperstidning* nr 181974, p. 681.
10. Brown, M.L., "Pollution Control Equipment," Georgia Institute of Technology, Engineering Experiment Station, Atlanta, GA.
11. Bulpitt, William S., et al, *A State Demonstration Program in Wood Energy—Volume II,* Georgia Institute of Technology, Engineering Experiment Station, Atlanta, GA, 1980.
12. Bulpitt, W.S., McGowan, T.F., and others, "Wood-Fueled Processes and Equipment," Georgia Institute of Technology, Engineering Experiment Station, Atlanta, GA, May 1980.
13. B.W. Associates, "Source Test Report—Rotary Chip Dryer," *Timber Products,* Medford Oregon-May 15, 1974," B.W. Associates, Klamath Falls, OR.
14. Cassens, Daniel L., and Choong, Elvin T., "Fuel Values of Southern Hartwood Mill Residues," *Southern Lumbermen,* November 1976.
15. Chermisinoff, Nicholas P., *Wood for Energy Production,* Ann Arbor Science, Ann Arbor, MI, 152 pp., 1980.
16. Clifton, David S., Jr., Bulpitt, William S., *A Feasibility Study for Wood Energy Utilization in Georgia,* Final Report, Project A-2140, prepared for The Georgia Forestry Commission under the sponsorship at The Coastal Plains Regional Commission, Georgia Institute of Technology, August 1979.

17. Committee on Industrial Ventilation, *Industrial Ventilation—A Manual of Recommended Practice,* American Conference of Government Hygienists, Lansing, MI, 1978.

18. "Compilation of Air Pollutant Emission Factors," U.S. Environmental Protection Agency, 1975.

19. *Cost Comparison Study Industrial Size Boilers 10,000 to 400,000 Pounds per Hour,* Davy McKee Corporation, Cleveland, OH, 44131, DOE Contract EX-77-C-01-2418, October 1979.

20. Curtis, G.B., "Hazards Associated with Industrial Wood Combustion," Georgia Institute of Technology, Engineering Experiment Station, Atlanta, GA, 1979.

21. Curtis, G.B., et al, *A State Demonstration Program in Wood Energy—Case Studies in Wood Energy,* Georgia Institute of Technology, Engineering Experiment Station, Atlanta, GA, October 1980.

22. Curtis, G.B., Brown, M.L., and others, "Wood Energy Economics—Proceedings April 30, 1980," Georgia Institute of Technology, Engineering Experiment Station, April 1980.

23. Curtis, Michael L., *The Effects of Outside Storage on the Fuel Potential of Green Hardwood Residues,* Thesis, Virginia Polytechnic Institute and State University, Blacksburg, VA, June 1980.

24. Federal Register, Thursday, September 13, 1979, Part IV, Criteria for Classification of Solid Waste Disposal Facilities and Practices: Final, Interim Final, and Proposed Regulations, 40 CFR Part 257.

25. Federal Register, Monday, December 18, 1978, Part IV, Hazardous Wastes, Proposed Guidelines and Regulations and Proposal on Identification and Listing, 40 CFR Part 250.

26. Feist, W.C., Springer, E.L., Hajny, G.J., "Viability of Parenchyma Cells in Stored Green Wood," *TAPPI,* Vol. **54,** no. 8, August 1971.

27. Fennely P.F., et al, "Environmental Assessment Perspectives," GCA Corp. for U.S. EPA, Washington, D.C., March 1976, p. 213.

28. Ferm, Bo. C., "Improved Bark Boiler Efficiency and Capacity by the Utilization of the Flue Gas for Pre-Drying," Bahco Systems Inc., Atlanta, GA, 1978.

29. Halsam, Robert J., and Russell, Robert P., *Fuels and Their Combustion,* McGraw-Hill Book Co., New York, NY, 1926.

30. Hedgecock, James S., and Dr. D.H. Steensen, "North Carolina's Wood and Bark Residues— Her Undervalued Assets," Misc. Ext. Pub. No. 83, School of Forest Resources, NC State University, P.O. Box 5488, Raleigh, NC, 27650, 71 pp., April 1972.

31. Hiser, Michael L., *Wood Energy,* Ann Arbor Science, Ann Arbor, MI, 151 pp., 1977.

32. Jahn, Larry G., *Directory of Residue Producers in North Carolina,* Mimeo, Extension Forest Resources Department, NC State University, P.O. Box 5488, Raleigh, NC 27650, 19 pp., July 1978.

33. King, W.W., *Survey of Sawmill Residues in East Texas,* Tech Rpt. No. 3, Forest Products Department, Texas Forest Service, Lufkin, TX, 59 pp., February 1952.

34. Koch, Peter, *Utilization of Southern Pines,* Ag. Handbook No. 420, USDA, Forest Service, SEFES. For sale by Supt. of Doc., U.S. Gov't Printing Office, Washington, D.C., 20402. Price $11.25 for two volume set. Vol. II, Processing, Chapter 29, "Measures and Yields of Products and Residues," 1663 pp., January 1972.

35. Kramer, John V., "A Hazard Management Protocol for Hazard-Control Dependability," Institute of Safety & Systems Management, University of Southern California, presented at the Western Occupational Health Conference, 10/11–15/79.

36. Kramer, Steve, *Health and Safety Requirements for Wood Handling Systems,* Georgia Institute of Technology, Engineering Experiment Station, Atlanta, GA.

37. Labosky, Peter, Jr., P.J. Przes Trzelski, and C.C. Rountree, Jr., "The Production and Utilization of Bark in South Carolina in 1975," Forestry Notes No. 18, Coop. Extension Service, Clemson University, SC, 15 pp., October 1976.

38. Levi, Michael, P., et al., *Decision Makers Guide to Wood Fuel for Small Industrial Energy Users*, National Technical Information Service, Springfield, VA, 127 pp., 1980.

39. Malte, P.C., Cox, R.W., Robertus, R.J., Messinger, G.R., Strickler, M.D., "Wood Particle Drying Rates," Eleventh Washington State University Symposium on Particleboard, Pullman, WA, March 1977.

40. McGowan, Thomas F., Walsh, James L., Jr., *Wood Fuel Processing*, Volume III, Project A-2400, U.S. Department of Energy Contract, DE-FG05-79ET23076, Georgia Institute of Technology, June 1980.

41. Miller, Bryan, *Safe Wood Storage*, Champion Papers, Courtland, AL.

42. "National Primary and Secondary Ambient Air Quality Standards," Federal Register 36, #84, pp. 8186–8201.

44. National Safety Council, 444 N. Michigan Avenue, Chicago, IL 60611, Industrial Data Sheets:
    Unit First Aid Kits, No. 202
    Log Skidding by Tractor, No. 377
    Tools for Manually Handling Logs and Pulpwood 292
    Poison Ivy, Poison Oak, and Poison Sumac No. 304
    Tick Bites, No. 228
    Grounding Portable Electrical Equipment No. 299
    Locking out Electrical Switches, No. 237B
    Portable Power Chain Saws, No. 320
    Brush Cutting Tools, No. 427-78
    Underground Belt Conveyors, No. 477
    Workplace Accident Records and Analysis, No. 527-B
    Roller Conveyor, No. 528A
    Belt Conveyors for Bulk Materials, Part I, Equipment No. 569
    Belt Conveyors for Bulk Materials, Part II, Operations No. 570
    Front End Loaders, No. 589
    Chippers and Hogs, No. 602
    Management Policy's on Safety and Health, No. 585
    Tractor Operation and Roll-over Protective Structures, No. 622
    Ground Fault Circuit Interruptors for Personnel Protection, No. 636
    Entry into Bins and Tanks, No. 663
    Writing & Publishing Employee Safety Regulations, No. 664

45. Nordfeldt, Sven., "The Improvement of Wood Waste Fuels Through Mill Drying," Flakt, Inc., Technical Bulletin, Vol. **3,** No. 4, December 1979.

46. *OSHA General Industry & Safety Health Standards*, 29 CFR 1910, 11/7/78.

47. Page, Rufus H., and Baxter, H.O., *A New Look at Residues from Georgia's Primary Wood Manufacturing Industries*, Georgia Forest Research Paper 78, Georgia Forest Research Council, Macon, GA, 11 pp., November 1974.

48. Perry, Joe D., and Gregory, R.P., *A Guide to 1975 Sawmill Residue Volumes in the 125 Tennessee Valley Counties*, Tech. Note No. B20, Div. of Forestry, Fisheries and Wildlife, TVA, Norris, TN, 55 pp., December 1976.

49. Perry, Robert H., and Chilton, Cecil, H., *Chemical Engineers' Handbook, Fifth Edition*, Ch. 9, McGraw-Hill Book Co., New York, NY 1973.

50. Phillips, Douglas R., "Lumber and Residue Yields for Black Oak Saw Logs in Western North Carolina," Forest Products Journal, Vol. **25,** No. 1, pp. 29–33, January 1975.

51. "Prevention of Significant Deterioration Standards," Federal Register, Vol. **42,** No. 212, 11/3/7//, p. 57459.

52. *Prevention of Significant Deterioration—Workshop Manual*, U.S. Environmental Protection Agency, Research Triangle Park, NC, October 1980.

53. "Public Law 91-596," 91st Congress S. 2193, 12/29/70.

54. The Resource Conservation and Recovery Act, Public Law 94-580, 94th Congress, October 21, 1976, as amended 1978.

55. Rules and Regulations for Solid Waste Management, Rules of Georgia Department of Natural Resources, Environmental Protection Division, Chapter 391-3-4, October 1974.

56. "Safety Standards for Conveyors & Related Equipment," B20.1, OSHA, Washington, D.C.

57. Safety Code for Cranes, Derricks and Hoists, Overhead and Gantry Cranes, B30.2.0-1967, OSHA, Washington, D.C.

58. Safety Code for Cranes, Derricks, and Hoists-Crawlers, Locomotive, and Truck Cranes, B30.5-1968, OSHA, Washington, D.C.

59. Saucier, Joseph R., *Environmental Effects of Harvesting Energy Wood,* U.S. Forest Service, Athens, GA.

60. Solid Waste Management Act, Act 1486, Georgia Laws of 1972 as amended through 1973.

61. Shirley, A. Ray, *Georgia's Wood Energy Program,* Georgia Forestry Commission, Macon, GA, 1979.

62. Springer, E.L., and Hajny, G.J., "Spontaneous Heating in Piled Wood Chips," *TAPPI,* Vol. **53,** No. 1, January 1970.

63. Springer E.L., Hajny, G.J., and Feist, W.C., "Spontaneous Heating in Piled Wood Chips—Effect of Temperature," *TAPPI,* Vol. **54,** No. 4, April 1971.

64. *Steam—Its Generation and Use,* 37th Edition, The Babcock and Wilcox Company, New York, NY, 1963.

65. Switzer, G.L., Nelson, L.E., and Hinesley, L.E., 1978, "Effects of Utilization on Nutrient Regimes and Site Productivity," In C.W. Millin (ed.)., *Proc. Complete Tree Utilization of Southern Pine,* New Orleans, LA, March 1978, pp. 91–102, Forest Products Research Society, Madison, WI.

66. Thompson, Stanley, P., "Fuel Preparation Systems Using a Rotary Dryer," Rader/Thompson Rotary Dryer, Rader Systems, Memphis, TN 1979.

67. Tinely, J. Lewis, *Ash Disposal Methods,* Georgia Department of Natural Resources, Environmental Protection Division, Atlanta, GA.

68. Tillman, David A., *Wood As An Energy Resource,* Academic Press, New York, NY, 252 pp., 1978.

69. Wells, Carol G., and Jorgensen, Jacques R., 1979, "Effects of Intensive Harvesting on Nutrient Supply and Sustained Productivity," In *Proc. Impact of Intensive Harvesting on Forest Nutrient Cycling,* p. 212–230, College of Environmental Science and Forestry, Syracuse, NY.

70. White, Marshal S., "The Effect of Storage on Wood Fuels," Proc. No. P-80-26, Forest Products Research Society—*Energy Generation and Cogeneration from Wood.*

71. White, Marshal S., and De Luca, Peter A., "Bulk Storage Effects on the Fuel Potential of Sawmill Residues," *Forest Products Journal,* Vol. **28,** No. 11, November 1978.

72. Williams, David L., and Hopkins, W.C., *Converting Factors for Southern Pine Products,* Bulletin No. 626, Louisiana State University, Ag. Exp. Station, Baton Rouge, LA, 89 pp., March 1969.

73. *Wood as an Industrial Fuel,* Georgia Institute of Technology, Engineering Experiment Station, Publication of conference on 10/31/79.

74. *Wood Energy Economics,* Proceedings, Georgia Institute of Technology, Atlanta, GA, April 30, 1980.

Appendix 5
Glossary

ACFM: Actual cubic feet per minute. The measured flow rate at process conditions as opposed to a flow rate that has been adjusted to standard temperature, pressure, and humidity.

Air Sweep Feeder: Device which uses air to transport fuel to the furnace and distribute it.

Auger: A screw encased in a tube used for moving material.

Baghouse: A chamber into which the boiler exhaust is directed. The chamber is fitted with fabric filters whose purpose is to collect solid material in the flue gas.

Belt conveyor: A device for transporting material, consisting of a flat continuous belt.

Boiler, HRT: Horizontal return tube boiler. A boiler in which the water to be vaporized is housed in a drum through which horizontal tubes (firetubes) are run. Gases from an external furnace are passed.

Boiler, radiant: A boiler in which the watertubes are usually touching the furnace walls and steam is generated largely by radiant heat (as opposed to heat from convection of gases about the tubes).

Bottom ash: That ash that remains in, on, or beneath the grates after burning.

Btu: An abbreviation for ''British thermal unit''—the amount of heat that is required to raise one pound of water one degree Fahrenheit.

Bucket elevator: A series of buckets connected by flexible lengths used to transport material—usually to lift material to a higher elevation.

Bulk density: The average density found in a large volume of material. Expressed in weight/unit volume (i.e. lb/cu ft).

Calcining: The heating of inorganic materials to a high temperature to drive off volatile matter or effect changes such as oxidation or pulverization.

Carbon monoxide (CO): A toxic gas resulting from incomplete combustion of carboniferous fuel.

Carbon, activated: A highly adsorbent powdered or granulated carbon.

Carbon, fixed: Carbon not driven off by heating to 1700°F.

Chain conveyor: A series of receptacles or buckets connected by links so as to form an endless belt whose purpose is to move material from one point to another.

Char: A combustible residue from organic material, i.e., charcoal.

Cogeneration: The simultaneous production of electricity and thermal energy.

Combustion thermal efficiency: The theoretical heat energy available from a given fuel compared to the actual heat energy made available by burning that fuel.

Condenser: A heat exchange device in which gas is changed to a liquid through the removal of heat.

Condensing steam turbine: A turbine in which steam exhaust is condensed.

Conveyor: A mechanical apparatus for carrying bulk material from place to place such as an endless moving belt or a chain of receptacles.

Cubes: Densified wood in a cube shape similar in size to some varieties of animal feeds.

Cyclone: A device for removing dust from gas by centrifugal action.

Cyclone burner: A device in which dry fine partices of wood are mixed with combustion air and blown tangentially through a series of manifolds into a cylindrical firebox.

Densification: A process whereby wood is mechanically compressed and the density of the material increased.

Dry scrubber: A device that traps the particulate matter in a gas in a moving bed of granular material. The trapped particulates may then be removed from the media by recycling.

Dryer, cascade: A device in which material is dried by falling through screens of hot gas.

Dryer, flash type: A device consisting of several loops of duct where wet material and hot flue gases mingle and drying occurs.

Dryer, flue gas: A dryer which utilizes the exhaust gases of a furnace or boiler.

Dryer, rotary: A rotating cylinder through which solids to be dried and hot gases are passed.

Dryer, tunnel: A unit in which solids are progressively dried by being moved through a tunnel in contact with hot gases.

Electrostatic precipitator: A device that ionizes the particles in the gas as they enter the unit and then traps the charged particles by attracting them to oppositely charged collection plates. The plates are cleaned periodically by mechanical rapping and the dislodged particles are then collected in a hopper.

Emissions (with reference to fuel combustion): The constituents which make up the total of the exhaust products.

Ethanol: Grain alcohol or "drinking" alcohol.

Fine: A very small particle of material such as very fine sands or very small pieces of bark.

Fluidized-bed combustion: A furnace whose combustion chamber floor has many fine holes through which underfire air is forced. This air blows through a bed of noncombustible materials such as sand or small particles of limestone. The air pressure and volume are such that the bed is kept in suspension. The bed is then heated to a temperature that will ignite the fuel to be burned and the burning of the fuel then maintains the bed temperature.

Fly ash: Fine solid particles of non-combustible ash carried out of a bed of solid fuel by the draft.

Fractionating scrubber, system: In this system, a manifold is attached to a cyclone below the tube sheet which allows a controlled extraction of the air from the large cyclone cavity.

Front end loader: A tractor with a bucket mounted on the front of the vehicle. The bucket is hydraulically actuated so that it may scoop up material, transport it, and then unload the material.

Furnace, rotary: A furnace in which the hearth is annular shaped and may be rotated a varying speeds.

Furnace, shaft: A vertical, refractory-lined cylinder in which a column of solids is maintained, and through which an ascending stream of hot gases is forced.

Gasification: The process of converting a carbonaceous material into (primarily) a gas.

Grate: A device to support the fuel while it burns.

Grate, dump: A grate used with a spreader stoker that can itself dump ashes into the ashpit.

Grate, inclined: A device that supports the fuel in other than a horizontal position, i.e., a sloping position.

Grate, traveling: A continuous cleaning grate that fits over bars which in turn are attached to chains that form an endless belt.

Hammermill: A device used to reduce the size of material by hammering action.

HC: Hydrocarbons.

Heating value, higher: A measure of heat energy of wood at any specified moisture content.

Heating value, lower: The higher heating values less the heat energy required to vaporize the moisture in the fuel.

Hogged material: Material (wood) which has been processed to a specific size, i.e., wood chips.

Live bottom (with reference to material handling equipment): A material storage bin whose floor incorporates a device for removing or unloading the material contained in the bin.

Magnetic separator: A device for removing iron and steel from other material through the use of magnetic attraction.

Mass: The amount of matter contained in a particle.

Material screen: A screen used to group materials according to their size. The openings in the screen have a specific size and material that is larger than the opening will not pass through the screen.

Metering bin: A device that both stores material and dispenses the material at a given rate.

Methanol: A toxic liquid commonly known as wood alcohol.

Multiclone: Many small diameter cyclone tubes arranged so that the gas is progressively cleaned as it passes through the unit.

MWe: Megawatt, electrical.

Non-condensing steam turbine: The turbine produces power by acting as a pressure reducer and the low pressure turbine exhaust becomes the process steam.

NO_x: Oxide of nitrogen.

Oil, catalytic: Oil produced by subjecting wood to high temperature and pressure in the presence of alkaline metal catalysts.

Oil, pyrolytic: Oil produced by low temperature, partial combustion of wood. The oil contains varying amounts of water, can be burned in modified heavy oil burners, and can be used as a chemical feedstock. Unlike fuel oils, pyrolytic oil is corrosive and contains acetic, formic, propionic, and other acids and must be stored and transported in corrosion resistant materials.

Opacity regulations: Regulations of the Environmental Protection Agency which judge the emissions from a furnace by comparing the opacity of the plume with a known standard.

Overfire air: Combustion air which is introduced above the fuel bed.

Particle resistivity: The measure of a particle's ability to accept and hold an electrical charge.

Particle strength: The physical strength of a particle, a property that must be considered when devising a fly ash collection system.

Particulates: Minute separate particles.

pH: The concentration of hydronium ions in solution. A method of expressing acidity or alkalinity on a scale whose values are from 0 to 14 with 7 representing neutrality, numbers less than 7 increasing acidity, and numbers greater than 7 increasing alkalinity.

Pile Burning: Combustion of the fuel while in mounds or piles.

Pneumatic conveyor: A device to transport material by using high velocity air.

Pre-drying: Removing the moisture from the wood prior to using it as a fuel, as opposed to letting the moisture be removed by vaporization as the wood is burning.

Precipitate: A substance separated from a solution or suspension by a chemical or physical change usually as an insoluble solid.

Proximate analysis: A statement of the volatiles, fixed carbon, and ash present in a fuel as a percentage of dry fuel weight.

Pyrolysis: A process of burning at less than stoichiometric conditions involving the physical and chemical decomposition of solid organic matter by the action of heat in the absence of oxygen. Products of pyrolysis may include liquids, gases, and a carbon char residue.

Radiant heat: Heat energy traveling as a wave motion (such as light travels).

Refractory lining: A ceramic lining capable of resisting (and maintaining) high temperatures.

Residence time: The length of time the fuel remains in a combustion zone.

Scrubber: An apparatus for removing impurities and contaminants from gases by use of water or a dry granular medium.

Shave-off scrubber system: This system utilizes a multiclone outlet incorporating two outlet tubes instead of one. The inner tube allows the core of cleanest air to pass through while the outer tube shaves off the perimeter of the outlet gases and recycles them in the cyclone.

Silicate: Metallic salt containing silicone and oxygen.

SO$_x$: Any of the various oxides of sulfur.

Specific weight: The weight of a substance as compared to some standard such as water.

Spreader-stoker: A stoker which throws or blows fuel into the firebox of a boiler so that it is more or less uniformly spread over the firebox grate.

Stoichiometric condition: That condition at which the proportion of the air-to-fuel is such that all combustible products will be completely burned with no oxygen remaining in the combustion air.

Stoker, underfed: A device which feeds the fuel to the fuel bed from below the point of air admission.

Suspension burner: A device to combust wood fuel which has been turbulently mixed with forced air in a stream over the main fuel bed.

Switchgear: Equipment for transferring electrical loads.

Thermocouple: Electrical device which measures temperature.

Turbine, back pressure: A turbine whose exhaust steam is used as process heat and the turbine work is considered, more or less, a by-product.

Turbine, extraction: A turbine from which partly expanded steam may be extracted for process heat. The turbine will usually sustain full rated output with or without extraction.

Turndown ratio: The lowest load at which a boiler will operate efficiently as compared to the boiler's maximum design load.

Turnkey System: A system which is built, engineered, and installed to the point of readiness for operation by the contractor.

Ultimate analysis: A description of a fuel's elemental composition as a percentage of the dry fuel weight.

Underfire air: Combustion air which is introduced below the fuel and rises through it.

Vibrating conveyor: A device that moves material by itself vibrating.

Volatiles: Substances that are readily vaporized.

Wet scrubber: An apparatus that develops an interface between a scrubbing liquid and the gas to be cleaned. Particles in the gas are trapped by liquid droplets and the liquid is then collected and removed.

Wood chipper: A mechanical apparatus for making wood chips.

Wood pellet: Processed wood that has been densified and shaped into the form of small cylinders very similar to animal feed.

Index

Boilers
  burners
    cyclone, 20. *See also* Manufacturers
    suspension, 20–21. *See also* Manufacturers
  dutch oven, 16
  field-erected, 14–15
  firetube, 13–14
  fluidized-bed. *See* Fluidized-bed combustors
  grates
    dump, 17
    rotary, 17
    sloping, 17
    traveling, 17, 19
  gravity feed problems, 17
  horsepower rating, 62
  load swing, 16
  manufacturers
    Bethlehem Corp., The, 25–27
    CNB Tri-Fuel Boiler, 27–28
    Deltak Corp., 30
    Industrial Boiler Co., 22–24
    Ray Burner Co., 23–25
    Weiss Boiler Co., 31–32
    Wellons, Inc., 28–29
  overfire air, 16
  packaged, 14–15
  performance, effects of moisture content on, 7
  pile burning, 15
  radiant heating, 17
  spreader stoker, 17–19
  systems. *See* Systems applications
  thermal inertia, 16
  turndown ratios, 17
  underfire air, 16
  watertube, 13–14
  wood consumption, 62
    quanity formula, 62

Chimney effect. *See* Storage of wood
Coal
  proximate analysis, 8
  systems. *See* Systems applications
Cogeneration
  cycles, 94–99
  economic considerations
    maintenance costs, 91
    operating costs, 98
    power costs, 96–97
    price, 91–98
  government incentives, 89
  introduction, 88
  steam turbines
    back pressures, 89
    condensing, 89
    extraction, 89
    maintenance, 90
    non-condensing, 89
  system size, 98–99

Drying of wood fuel
  economics, 84
  equipment
    cascade dryers, 81–82
    flash dryers, 81–82
    rotary dryers, 80–81
  methods, 79
  new installation, 84
  retrofits, 82–84

Economic analysis of wood energy systems
  cash flow analysis, 147–151
  depreciation, 146
  discounted payback period, 151
  discounted rates of return on investment, 151
  investment tax credit, 146
  sensitivity analysis, 151
  simple payback period, 151, 165
  tax considerations, 146
  turnkey, 145
Economics
  capital costs, 162
  case study. *See* Integrated Products, Inc., case
    study
  fuel costs, 162
  steam costs, 162–165
Emissions from boilers
  baghouse, 20
  control techniques
    baghouses, 109–110
    collection efficiencies of control devices,
      107
    costs, 114–118
    dry scrubbers, 112
    electrostatic precipitators, 112–114
    mechanical collectors, 108–109
    wet scrubbers, 109–112
  factors affecting emissions
    boiler design, 103
    boiler operation, 104
    emission characteristics, 106
    emission factors, 105
    fuel, 103
  flue gas desulfurization, 20
  fly ash and particulate, 101, 107
  regulations
    Environmental Protection Agency, 102
    Georgia, 101–102
  types of emissions, 101
  1976 Clean Air Act, 100
Environmental impact, pertinent regulations
  National Ambient Air Quality Standards, 120
  New Sparse Performance Standards, 123–124
  Prevention of Significant Deterioration—Clean
    Air Act 1977, 120–123
  Solid Waste Management Act, 124–126

Feasibility study methodology
  order of investigation, 134
  requirements

economic analysis. *See* Economic analysis
    of wood energy systems
  energy load cycles, 134–137
  existing data, 138
  facility layout, 139
  installation planning, 140
  operational, 140
  system and fuel, 137
Fluidized-bed combustors, 38–45
  systems. *See* Systems applications
Fuels derived from wood
  costs, 155
  introduction, 152–153
  types, 154

Gas boiler systems. *See* Systems applications
Gasifiers, 48–56

Handling of wood fuel
  methods
    augers, 75
    belt conveyors, 74
    drag chains, 75
    front end loaders, 75–76
    others, 76
  problems
    bridging, 73
    ratholing, 73

Integrated Products, Inc., case study
  background data, 172–175
  boiler system description, 180–181
  cash flow summary, 185
  contractor selection, 177–179
  costs, 182–183
  economic analysis, 181–182
  life cycle cost analysis, 183–184
  performance report, 184–186
  purchase contract content, 178–180
  storage and handling system, 174–175
  wood fuel availability analysis, 184

Manufacturers
  cyclone burners
    Coen Co., 34, 37
    Guaranty Performance Co., 34, 36
    McConnell Industries, 34
    Moore-Oregon Corp. and Energy Ltd., 32–
      34
  fluidized-bed combustors

Combustion Power Co., 44–45
Johnston Boiler Co., 41, 43–44
York-Shipley Inc., 41, 42
pyrolysis equipment
Energy Resources Co., Inc., 47–48
Tech-Air Corp., 45–47
suspension burners
Peabody Gordon-Piatt, 38
wet cell
Lamb-Cargate Co., 57–59
wood gasifiers
Forest Fuels Co., The, 54
Moteurs Duvant, 54–56
University of California/Davis, 51–52, 54

Oil boiler system. *See* Systems applications

Pellets. *See* Wood pellets
Preparation of wood fuel
drying. *See* Drying of wood fuel
size reduction
disk screen, 77–78
equipment costs, 78
hammermill, 77–78
hog, 77
metal separator, 77–78
Processing network, 169–171
Properties of wood fuels
approximate weights, 11
ash contents, 7
available energy, 66–67
bulk density, 61
combustion
stages, 10–12
temperature, 10–12
dry basis, 4–5
extruded wood. *See* Wood pellets
flame temperature, 78
heating value
higher, 8–10
lower, 9–10
net, 9, 61
per cord, 11
higher heat value, 6
moisture content, 3–5, 70
particle size, 6
pH, 68
proximate analysis, 6
silica content, 61
size, 60
sources, 60–61
standards, 61
ultimate analysis of wood species, 8
wet basis, 4–5, 61
Purchasing of wood fuels, 143–144
Pyrolysis, 45–48

Sand. *See* Properties of wood fuels, silica content
Safety
boiler flame safeguard, 20
National Fire Codes, 128–129
OSHA, 126–128
Storage of wood fuel
bin, 66
chimney effect, 67–68
construction, 68
covered, advantages of, 71–72, 84–87
effect of moisture, 66–68, 70
fire hazards, 70
open, 65–71, 84–87
packing density, 66–67
pile temperatures, 70
pilers, 66–68
procedures
sawdust and mill residue, 69
whole green chips and/or bark, 69
silos
costs, 73
sizing, 72–74
sizing considerations, 65
thermocouple, 69
Supply of wood fuel, 141–143
Systems applications
coal, 159–161
design factors, 156
gas, 161
oil, 161
processes
capital cost vs. system capacity, 160
fluidized-bed, 159
fuel cost vs. system capacity, 160
waste wood system costs, 158
wood chip boiler, 158–159
wood pellet boiler, 159
steam generator, 155–162
wood/gas/oil, 161–162

Transportation and unloading systems
dump truck, 63–64

Transportation and unloading systems (*Cont.*)
  front end loader, 63
  labor, 63
  live bottom, 63–64
  loading time, 63
  railroad, 63–64
  scoop roveyor, 63, 65
  truck dump, 63, 64

Wood pellets, 61, 84
  systems. *See* Systems applications